Understanding Complexity

Scott E. Page, Ph.D.

THE
GREAT
COURSES

PUBLISHED BY:

THE GREAT COURSES
Corporate Headquarters
4840 Westfields Boulevard, Suite 500
Chantilly, Virginia 20151-2299
Phone: 1-800-832-2412
Fax: 703-378-3819
www.thegreatcourses.com

Scott E. Page, Ph.D.

Leonid Hurwicz Collegiate Professor
of Political Science, Complex Systems,
and Economics, University of Michigan
External Faculty Member, Santa Fe Institute

Professor Scott E. Page received a B.A. in Mathematics from the University of Michigan and an M.A. in Mathematics from the University of Wisconsin–Madison. He then received his M.S. in Business and his Ph.D. in Managerial Economics and Decision Sciences from the J. L. Kellogg School of Management at Northwestern University. He completed his Ph.D. thesis under the guidance of Stan Reiter and Nobel Laureate Roger Myerson. He has been a Professor of Economics at the Caltech and the University of Iowa and is currently Leonid Hurwicz Collegiate Professor of Political Science, Complex Systems, and Economics at the University of Michigan as well as a senior research scientist at the Institute for Social Research, a senior fellow in the Michigan Society of Fellows, and associate director of the Center for the Study of Complex Systems.

While a graduate student, Professor Page began visiting the Santa Fe Institute (SFI), an interdisciplinary think tank devoted to the study of complexity. He has been actively involved at SFI for more than 15 years. Currently, Professor Page serves as an external faculty member of SFI. For a dozen years, he, along with John Miller, has run a summer workshop for graduate students on computational modeling.

A popular advisor and instructor, Professor Page has won outstanding teaching assistant awards at the University of Wisconsin and Northwestern University, the Faculty Teaching Award at Caltech, and the Faculty Achievement Award for outstanding research, teaching, and service at the University of Michigan.

Professor Page's research interests span a wide range of disciplines. He has published papers in leading journals in economics, political science, ecology,

i

physics, management, public health, and computer science. He has served on dissertation committees for students in more than 10 departments. In recent years, his core interest has been the various roles of diversity in complex adaptive systems, such as economies and ecosystems. He is the author of two books on these topics, *Complex Adaptive Systems* (with John Miller) and *The Difference: How the Power of Diversity Creates Better Firms, Groups, Schools, and Societies*. Both books were published by Princeton University Press.

Professor Page has spoken on complexity and diversity to many leading companies, universities, and nonprofit organizations, including the World Bank, the Kellogg Foundation, Yahoo!, and the National Academies. He lives with his wife and two sons in Ann Arbor, Michigan. ■

Table of Contents

Table of Contents

SUPPLEMENTAL MATERIAL

Understanding Complexity

Scope:

Complexity science has become a phenomenon. Newspapers, magazines, and books introducing the core concepts from complexity science (emergence, tipping points, the wisdom of crowds, power laws, scale-free networks, and six degrees of separation, to name just a few) have flooded the mainstream. This popularization has taken place at the same time that complex systems techniques have gained a foothold in the academy. The analyses of political systems, economies, ecologies, and epidemics increasingly invoke concepts and techniques from complexity science.

In this course, we learn the nuts and bolts of complexity. We cover the core concepts and ideas that have transformed complexity from a loose collection of metaphorical intuitions into a respected scientific discipline in less than a quarter century. The science of complexity contains a plethora of models and ideas through which we can interpret and understand the world around us. Thus we keep an eye on reality throughout, touching on the practical benefits of gaining an understanding of complex systems.

We begin by defining complexity, which proves to be rather challenging. Many people confuse complexity with chaos, difficulty, or complicatedness; it's none of the three. We'll see that different disciplines rely on distinct conceptions and measures of complexity. What a computer scientist and an ecologist call complexity do not align perfectly, but the definitions will be close enough for us to get some leverage.

One of our goals will be to see complexity science in a transdisciplinary light, as a set of ideas, concepts, and tools that can be applied across disciplines. Hence in the lectures that follow, we flow in and out of traditional disciplinary boundaries. We will discuss the spread of diseases, the collapse of ecosystems, and the growth of the Internet. Sometimes we will link concepts tightly to specific real-world situations. Other times we will paint with a broad brush and discuss how a core concept transcends boundaries.

One of those transcendent ideas is the notion of emergence. Emergent macro-level properties arise through the interaction of lower-level entities and often bear no resemblance to them. The wetness of water is an emergent property, as is a heartbeat. We also discuss how in some cases, interactions produce critical states, in which small events can produce cascading changes. This process of self-organized criticality has been offered as an explanation for widespread power outages, stock market crashes, and traffic jams. Self-organized criticality is just one way in which small events can produce large disruptions. Systems can tip from one absorbing state to another. We'll learn why and how.

By definition, the topic of complexity is, well, complex. We cannot get around that. Throughout the course, we'll balance the need for precise understanding against the necessity of technical concepts. Jargon will be outlawed. The guiding principles for these lectures will be to move beyond metaphor (a butterfly flapping its wings) and anecdote (someone buying a pair of Hush Puppies) to a deeper, logical understanding of core concepts. Among the questions we take up: What is path dependence? What is a power law? How do you explain the phenomenon of six degrees of separation? What really is a tipping point? In addition, we'll learn to apply complex systems thinking to our personal and professional lives. Along the way, we'll have loads of fun! ■

Complexity—What Is It? Why Does It Matter?
Lecture 1

What are complex systems? What is complexity? ... And why does it matter?

In everyday life, the word "complex" might be applied to our lives, the economy, or even salad dressing. Our first step in constructing a science of complexity will be to construct definitions and measures. A system will be said to be complex if the whole transcends the parts. Most complex systems consist of diverse entities that interact both in space (either real or virtual) and in time. Most ecosystems, New York City, and an elementary school playground each satisfy this definition of complex, but a window air conditioner does not. Addressing many of our most pressing challenges—such as managing ecosystems and economies, or preventing mass epidemics and market crashes—will require understanding the functioning of complex systems.

When we describe something as complex, we mean that it consists of interdependent, diverse entities, and we assume that those entities adapt—that they respond to their local and global environments. Complex systems interest us for several reasons. They are often unpredictable, they can produce large events, and they can withstand substantial trauma. Complex systems produce bottom-up emergent phenomena, where what occurs on the macro level differs in kind from what we see in the parts. Emergence takes many forms, including self-organization. Finally, complex systems produce amazing novelty, from sea slugs to laser printers.

From a pragmatic perspective, our world forces us to take an interest in complexity. How do we make sense of it? In this course, we will formulate a reasonably precise definition of complexity so that we can accurately compare the complexity of situations. We will learn about complexity theory, a new way of thinking with new computational tools. We will move beyond metaphor and introduce the science and vocabulary of complex systems. A complex system is capable of producing structures and patterns from the bottom up. It does not settle into a simple pattern but instead is a source of near-perpetual novelty. It is not in equilibrium, nor is it chaotic.

A system can be considered complex if its agents meet four qualifications: diversity, connection, interdependence, and adaptation. Rather than just saying "economies are complex" or "ecosystems are complex," we can now base those statements on logical foundations.These four qualifications can also be used to prove that the world is becoming more complex by almost any measure: social, economic, political, physical, ecological, or biological. In addition, science has given us finer lenses with which to recognize complexity.

Our world forces us to take an interest in complexity.

At this point, we need to make an important distinction: Complex is not the same thing as complicated. Complicated systems may have diverse parts, but they are not adaptive. In addition to adaptability and robustness, complex systems have the ability to produce large events. Because they can produce large events, complex systems are often said to be not normal, or outside of the common distribution curve. It may seem paradoxical that complex systems are both robust and subject to large events, yet it is not.

Many people, including myself, study complex systems because of the mystery of emergence. Emergence is when the macro differs from the micro—not just in scale but in kind. One common form of emergence is self-organization. This occurs when a spatial pattern or structure emerges, such as flocks of birds or schools of fish. One fascinating thing about emergent phenomena is that they arise from the bottom up, without superimposed formalism.

Complex systems produce interesting dynamics such as phase transitions, which are sometimes called tipping points. Phase transitions occur when forces within a system reach what is called the critical threshold. Once this happens, the state of the system changes, often drastically. A phase transition is a type of nonlinearity. Unfortunately, there are many ways to be nonlinear. ■

Suggested Reading

Ball, *Critical Mass.*

Miller and Page, *Complex Adaptive Systems.*

Resnick, *Turtles, Termites, and Traffic Jams.*

Questions to Consider

1. How might you make more formal a claim that financial markets or international politics have become more complex?

2. Is making a situation less complex necessarily better?

Complexity—What Is It? Why Does It Matter?
Lecture 1—Transcript

Hi, my name is Scott Page. I received my training in mathematics and in economics, but over the past 10 years, I've worked as a professor of complex systems at the University of Michigan. People often ask me: What are complex systems? What is complexity? In this series of lectures, what we're going to do is take on this topic of complexity. We're going to ask: What is it? Why does it matter?

The word complex is tossed about in a variety of settings: on television and in newspapers, you might hear pundits describe financial markets or international politics as being complex. You might pick up the science news and read about medical researchers discussing complex interactions involved in the folding of a protein. Or you might be out for dinner and a chef might use the word "complexity" to describe the bay scallops served with Applewood bacon in a port reduction. But what do they mean exactly, and do they all mean the same thing?

In brief, when we describe something as complex, what we're going to mean is that it consists of interdependent, diverse entities; and in addition, we're going to assume that those entities adapt: that they respond to their local and global environment in some way. Often, it's going to be helpful to think of the entities arranged on a network or in some sort of space. So we have this definition—diverse, interdependent, connected, adapting entities—and if we take that we can see that yes, in fact, financial markets are complex. The banks, traders, government regulators are all connected in some sort of network, and their behaviors are interdependent: how one bank behaves depends on what other banks are doing. It's also the case that these banks adapt; unfortunately, in some cases they all adapt in the same direction, and we get bubbles and crashes. It's also true, then, that political systems are complex, right? How one country acts influences how another one does. As for those bay scallops, well, they're close; there's not a lot of adaption going on. But if they taste good enough, maybe we'll cut the chef a little slack.

Complex systems interest us for a bunch of reasons: First of all, they're often really unpredictable (that's not a very good thing; second of all, they often produce large events (we get crashes and wars, and that's also not a very good thing); but one good thing about complex systems is that they're remarkably robust, so complex systems can withstand unbelievable trauma and environmental variation and yet still hold together. We see this in ecosystems: They can lose a species, yet somehow maintain functionality.

Perhaps most interesting, though, is that complex systems produce what we call bottom-up emergent phenomena. This is where what happens at the macro level differs in kind from what we see at the micro level. Consciousness is an example of emergence: We have individual neurons that are really very, very simple things; yet we have a network of interdependent adapting neurons in our head that produce a conscious mind, a personality. This is amazing; emergence takes many forms, including self organization. The next time you're in a city, just step back and watch the comings and goings of people and delivery trucks and ideas and goods, somehow it all works. Yet there's no central planner; the order of a city just emerges from the bottom up.

Finally, when we look at a complex system, we see that they produce amazing novelty: toucans, sea slugs, laser printers, and hip hop music. Who could have imagined any of these things? It's these reasons—unpredictability, large events, robustness, emergence, and novelty—that we find complex systems interesting from an intellectual perspective.

From a pragmatic perspective, we have no choice but to take an interest in complexity. Like it or not, these are complex times. We live with market volatility, industry failures, warnings of ecological destruction, and terrorist threats; and there's little or no sign of this complexity waning. Be it not a wish or a curse then, but an inescapable reality: We live in complex times, so how do we make sense of them?

In this first lecture, we're going to stick to the basics; so most of what I want to do is I want to convince you that rather than think of "complexity" as some vague term, that we can give it a precise definition that will allow us to say that one situation is more complex than another. Remember, a complex system consists of these connected, interacting, diverse, adaptive entities: neurons, people, species, countries, firms. These parts are interdependent;

it's this interdependency and this adaptation that produce the complexity that is both our blessing and our burden.

In a series of lectures, we want to learn about complexity theory, and we're going to see this sort of as a new paradigm; as a new way of thinking. As often is the case, new paradigms bring with them new tools; and what we're going to see is that the new tools we use in this course are mostly computational. In the past, science was typically done through mathematics; now a lot of what we're going to see is computational tools put into play.

In these lectures, we're going to move beyond metaphor and we're going to introduce the science of complex systems. We're going to learn a whole new vocabulary: rugged landscapes, tipping points, positive feedbacks, emergence, self-organized criticality, and agent-based models. These are going to be new words that we learn, and we're going to use that vocabulary to make sense of the world.

Courses in complex systems tend to take one of two forms: If they're taught by a physicist, the course tends to rely entirely on precise logical definitions, mathematical theorems, and formal models. If they're taught by a social scientist (like me), or a journalist even, the course would rely on loose definitions, metaphors, and anecdotes. We're going to shoot for sort of a middle ground here. We're going to introduce formal definitions when needed, but we're not going to let formalism bog us down; we're not going to let it get in the way of a deep conceptual understanding of the core insights. To get that deep conceptual understanding, we're going to talk about anecdotes and historical events.

So this opening lecture, I just have three goals: First, we're going to learn what we mean by complexity science, to give us some formal definitions; second, to see why complexity is important, why it matters; and then third and finally, I want to just give some glimpse of what's going to come throughout the series of this lecture since most of us probably don't know what complexity science is. So let's get to it; let's go.

What is a complex system? A complex system is capable of producing structures and patterns, and it's these high level structures and patterns that just emerge from the bottom up, that's sort of amazing; they're not built in. Life itself is an emergent phenomenon. Complex systems don't settle into simple patterns—little periodic orbits, like the earth going around the sun—but instead they're a source of constant novelty. Again, as I mentioned before, they're paradoxically both robust and susceptible to large events; so a small or moderate disturbance can have little or no impact, yet at the same time a small grievance can escalate into a large war. For that reason, we're sometimes going to describe throughout this course complex systems as not normal. By that I mean that outcomes don't fall on a standard normal distribution—the standard sort of bell curve that we see in statistics—but instead, they're distributed according to something called a power law. Power law is a distribution where we get these very large events; I'll explain that in much more detail in a later lecture.

What's interesting about complex systems—at least to me—is that they pulse with life. They're not in some sort of equilibrium—so we don't have supply equaling demand like an economist would think—nor are they chaotic, but they occupy this strange place in between order on the one hand and chaos on the other. So let's dig deeper into our definition: We're going to define a system as complex if it consists of diverse agents who are connected whose behaviors and actions are interdependent and who adapt. These four parts—diverse, connected, interdependent, and adaptive—are the necessary parts for a complex system. So rather than just saying "economies are complex" or "ecosystems are complex" or "the international political scene is complex," we can base those statements on logical foundations.

I want to take these one by one. An economy is diverse; it has diverse producers: there's McDonalds, there's Williams-Sonoma. An ecology has diverse parts: there are houseflies, squirrels, and (in my backyard) there are Rose-breasted Grosbeaks.

Both economies and ecosystems are also connected; so in an economy, firms are connected to suppliers in a supply network. Two grocery stores that are in the same town are also connected geographically. Or if I go online, even, two water ski suppliers are connected virtually. Again, we have all these connections in an economy. In an ecology, species typically are connected by geography: two species interact if they happen to be in the same location.

These same species are what we call interdependent; and what I mean by this is nature red in tooth and claw: Species literally feed on one another. There are also interdependencies in an economy; at least on the surface these seem a little bit less severe—there's no eating and being eaten—but there is competition: how one firm does determines how another one does. In the social realm, we also have interdependencies. These are more psychologically based; so how happy I am depends on other people think about our choices. If I'm wearing some purple Converse All Stars and people like them, I feel happy; if they don't like them I feel less happy.

Last but not least: adaptation and selection. Let's take a company like McDonalds. They offer salads; when they do this, they're adapting to a changing environment. Markets also have selection; it wipes out poor performers. Witness the failure of Montgomery Ward or Oldsmobile. Ecosystems also feature this adaptation and selection. With even a crude observation of the world around us, we can see adaptation in practice. I was in Palo Alto for a year, and acorns were scarce. What happened was the squirrels started eating the soy-based electric wires in my car; that's adaptation. Selection in ecosystems is more difficult to see in real time, but the fossil record makes clear that those species that do not fit in—those that are not fit—don't survive.

Now that we have these four attributes—diversity, interdependence, connectedness, and adaptation—we can put some flesh on my earlier claim that the world is in fact becoming more complex. Let's compare our world today against that of an earlier time. 10,000 years ago, the earth supported 5 to 10 million people—that's it—and those people were mostly concerned with survival; they weren't concerned with watching movies. They hunted animals; they gathered berries. At recent as 100 years ago, most of the world's population lived in rural areas and worked by farming or hunting and gathering. Farming is difficult work, don't get me wrong; you pick rocks in the spring, you plow fields, you plant, you milk and birth cows, etc. The average farmer, though, belonged to a community of people who looked like them; there wasn't much diversity. Those other farmers had similar skill sets; those farmers worked for the most part in isolation. Maybe once a month the farmer would head into town to get sugar, cloth, sell some crops. So you have not much diversity, not much interdependence, and not much adaptation.

As the years passed the door on a typical farm, each one went roughly the same as the one before: you plant, weed, harvest. Yeah, there were variations in weather; but the key point here is that if the farmer planted beans and carrots, he got beans and carrots. Nothing unpredictable happened.

Let's compare this to the modern world. Now what most workers do is much more diverse; people live in cities, they're more connected. In developed countries, they have incredibly diverse skills; just look through the Yellow Pages: you find engineers, physicists, lawyers, psychologists, landscape architects, computer repair people, and so on. These people are more connected as well. These connections are both physical (we live in closer proximity and more and more work is also done in teams) and life is also virtual now (we have friends we connect to online and by phone); so we're just much more connected.

It's also true that our actions are more interdependent. When we work in a group on projects or a team, the value of one person's contribution depends on the values of others. Whether we're making a movie, a piece of software, or a car, the success of that project depends on the coordinated actions of many people. The farmer of old reaped what he sowed. The modern knowledge worker is only as good as the team that surrounds her and the context in which she finds herself. We're much more interdependent.

We're also more adaptive. New firms and technologies are constantly arriving. We have better and faster information. Prior to what's been called the information revolution, if you ran a large company, you would wait six to nine months to see how sales were going. That's completely changed; now firms get minute by minute updates. Let me give a very specific example: If you go into a major store like a Wal-Mart and you buy milk, not only do they update their inventories; they can place new orders for more milk automatically if supplies warrant it. Let's suppose you're in Peoria, Illinois, and you stop and you put some milk in your cart and you walk to the checkout. It could very well be the case that a refrigerated truck driving down I-80 gets sent a message and just turns right back around and heads back to Peoria and restocks those shelves. It's been said of Wal-Mart in particular that as soon as you place milk in your cart they call up the cow.

In painting this broad-stroke picture of increasing complexity—especially the increased adaptation that we saw in the case of Wal-Mart—three caveats are in order: First, the increase in diversity, connectedness, interdependency, and adaptivity doesn't mean that the world was not complex in the past. It was; it just wasn't as complex as the world is now. Second, this argument that I've just made for increased complexity doesn't make any value judgments about the past relative to today. I'm not saying that life in the past was boring; I'm not saying it was uninteresting; all I'm saying is that technological advances and increased population have ramped up the attributes that contribute to complexity. Third, we may not even be saying that the world will keep getting more complex; we're going to talk about this in a later lecture. It's possible the world could become less complex. If we ramp up these attributes too much—if we become too interdependent, too connected, too diverse—we might lose complexity and what we might get instead is some sort of collapse into what's really just an incomprehensible mangle.

These caveats notwithstanding, by almost any measure the social, economic, and political worlds in which we live are more complex than they were a few hundred years ago. And they may be too complex for us to understand unless we develop some new ways of thinking; this is why I moved from economics into complex systems, because I felt the world was becoming more complex. It's also true—I've been talking about the political, social, and economic worlds—that physical, ecological, and biological worlds are also more complex. That seems impossible because physics and biology haven't changed; but think about it this way: Transportation has brought people and species in closer contact, so diseases, flora, and fauna now jump across physical space at speeds that were once impossible. We're also able to manufacture new physical entities: element 108, Hassium, didn't exist until man created it; so we have more things than we had in the past.

In addition, we have finer lenses with which to see the world and we're more aware of the complexity within us. This is especially true in biology. We used to think that perhaps there was maybe a clean mapping from DNA to phenotype, from genotype to phenotype. But now we're much more aware that there's an incredibly complex process that involves RNA and epigenetic processes (things outside of genes; chemical processes). The deeper we're able to look, the more complex the world around us and within us appears to be.

At this point, I want to stop a second and take a time out and make what I think is an incredibly important distinction: Complexity is not the same thing as complicated. A watch (I have a watch on my wrist here): A watch is complicated, but it's not complex. A watch has diverse parts—that's true—they're connected, and they're interdependent. But the key thing is those parts aren't adaptive, nor are they subject to selection. They're just mechanical parts that operate in a fixed manner. A watch can't do the things that complex systems do. It's not unpredictable, it's very predictable; it doesn't create large events, there's no large events coming from my watch; and it's not robust, it won't withstand small disturbances.

What do I mean by this? If I take off the back of this watch and pull out apart, it's going to break. That's because the other parts don't adapt; they're not going to fill in the gap that's left when I pull this part out. That's not true about organizations, and this is what's so cool about complex systems. Supposed you miss a day of work, or I miss a day of work. You come back the next day and your fellow employees say, "How are you feeling," "Is everything ok". They might even say, "The sledding was tough without you"; but the reality is the organization soldiered on without you just fine. Its robustness—the reason it could—is because the parts (the people) were able to adapt. Robustness is just one property of complex systems. Let's consider some others.

In addition to being robust, complex systems are going to produce large events. When I say large events, I mean large events: I mean floods, earthquakes, hurricanes; I mean things like political upheaval, stock market crashes, traffic jams; and I even mean biological large events like epidemics. Because complex systems can produce these large events, as I said before, we're going to sometimes say that they're not normal. Again, what I mean by that is the distribution of event sizes is not our sort of standard normal bell curve. Instead it has what we call a long tail; that means that there are these big events, and these big events have huge consequences—this is the world wars, the floods, the stock market crashes—that we really want to avoid. This seems paradoxical—or it should seem paradoxical   on the one hand, complex systems are robust, and on the other hand they're susceptible to these large events. I want to argue that this isn't paradoxical. Stock markets can crash 20 percent, cities can go underwater, and yet the economy and society can somehow muddle through; it's robust to these sort of shocks.

Let me give a very specific example from my own life. In the mid-1990s, I took my first job as a professor at Cal Tech in Pasadena, California. During my first year there, on January 17, 1994, at about 4:30 in the morning I was shaken from my bed by what has now been called the Northridge earthquake. During this earthquake the entire Los Angeles basin rattled and shook, and it seemed like minutes or a half hour but it was really just a few seconds. The impact of this earthquake was substantial. Four freeways were shut down because bridges just collapsed; and if you looked at aerial shots from the news helicopters, it looked really much closer to strategic bombing than some sort of natural event. And yet what's amazing is the Los Angeles economy survived with really just a small hiccup. How did it do it? People adapted. They telecommuted; they worked on adjusted schedules; some people left work early, some people left late. People were able to adapt and make the system robust.

It should be clear why from a purely pragmatic perspective—intellectually interesting as these ideas are—why this sort of paradox of robustness and large events should be of great concern to us. If we can discern how to construct robust systems—organizations, political systems, ecosystems—we'll be better off. At the same time, if we can identify conditions that are ripe for large events, then maybe we can head them off; maybe we can stem the tide. We'd also be better off.

I don't want to in any way to diminish these pragmatic considerations—I'm a social scientist and I'm sort of paid to be pragmatic—but they're really not the reason most people, including me, study complex systems. The reason we study complex systems is really the mystery; and one great case of mystery is emergence. So what is emergence? Emergence, as I said before, is this distinction where the macro differs from the micro. Consciousness is the classic example of emergence; we'll talk about that quite a bit throughout this course as an example.

I want to step back and sort of do some baby steps here. Thomas Schelling, who won a Nobel Prize in economics, wrote a book called *Micro Motives and Macro Behavior*. In this book, he worked through some examples, and in these examples what happens at the macro level doesn't accord with what seems to be going on at the micro level. Let me give an example: Suppose you have a bunch of people who are fairly racially tolerant and they're trying to decide where to live. What Schelling showed is that you'll end up with macro level segregation; so at the macro level you'll get segregation, even though at the micro level you have a fairly high degree of tolerance. This is a case where micro and macro disconnect writ large.

When we say they disconnected, I'm not saying that the whole just differs from the parts; I want to say that it actually differs in kind. What do I mean by that? Look at a single water molecule: There's no way in which a single water molecule can be wet. Wetness is a property that happens sort of at the macro level; it's something new and wonderful that emerges when one water molecule is combined with others. If we want to have any hope of understanding how wetness or consciousness emerges, we have to dig into the study of complex systems.

A common form of emergence is self-organization. This occurs when we get some sort of spatial pattern or structure: flocking of birds, schooling of fish, or in physics, crystal structures are what we call self-organized phenomena. What's cool about self-organization and other emergent phenomena is again, they arise from the bottom up; there's no central control. If you see a marching band in formation, that's top down organization; they're all getting marching orders, and they go exactly where they were told to go. If you see geese flying in a "V," you're seeing a structure that emerged without any sort of top down organization. You can spend all day in the park and you're never going to see the geese sitting in a little huddle where one person's telling the other geese where to go. It's also the case that water doesn't say, "Hey, let's all be wet." Your neurons don't have little committee meetings to decide how to remember the quadratic formula: $-b \pm \sqrt{b^2 - 4ac} / 2a$. There's no meeting that decides that, yet somehow my brain remembers that. It's these emergent phenomena that complex systems can shed light on, and in some cases—we hope anyway—explain.

The next thing I want to talk about is how complex systems produce something interesting from a dynamic perspective called phase transitions. These are more popularly known as tips or tipping points. Permit me an analogy here, and then an example. Imagine I've got a ball in the bottom of one basin of a double basin sink; so I've got a double basin sink here and the ball's in the bottom of one. Now I push that little ball up. What's going to happen is it's going to roll back down into the basis. Suppose I push a little harder, what's going to happen? It's going to roll back down in the basin. Let's suppose I give it a really hard push. Now it may escape the first basin and roll into the second basin. The speed at which that ball can escape is called the critical threshold or the escape velocity. The change in where the ball happens to be—the move from one basin to the next—can be thought of as a phase transition; we move from one state to another.

That was an analogy, let's get real here for a second; let's take this analogy and see it in action. Suppose I start with a clean, oligotrophic lake; a crystal-clear blue lake. I'm going to start dropping some phosphorous into this lake; maybe it's running off the fertilizer I've used. No big deal: What's going to happen is the ecosystem is just going to respond; the bottom sediments in this lake are going to absorb the increase in phosphorous. Let me drop in a little bit more. Again, no big deal, the lake is still clear; it's just like that ball going to the bottom of the basin. But suppose I drop in a little bit more? All of a sudden—boom—I'm going to get this big, murky eutrophic lake, a green mess with algae everywhere; because what's happening is the amount of phosphorous is going to cross this critical threshold, and we're going to get this phase transition from clear blue waters to green and algae-ridden.

Let me give another example, and this is a little more subtle. Let's think about increasing spending on education, healthcare, or poverty reduction. It could be that we have tipping points there as well. So maybe if we spend small amounts of money we have no effect, and if we spend a little bit more we also have no effect. But if there's a critical point—if there's a phase transition—it could be the case that if we spend a little bit more, suddenly everything will get a lot better. To know if that's true—to know if there is a phase transition, or critical or tipping point—what we need to do is we need to do a science of complex systems to understand how these systems work.

A phase transition is what we call a type of nonlinearity. Recall a linear function is a straight line; so a firm might find, for example, that their market share increases linearly with price decreases. If they cut prices by 1 percent, they get a 2 percent increase in sales. If they cut prices by 2 percent, they get a 4 percent increase in sales. In those examples, what you're seeing is a linear response. In the examples we just talked about, the response was nonlinear: We got nothing, nothing, nothing, and then boom. There's a problem here: There are a lot of ways to be nonlinear. In fact, Jon Von Neumann once said that the study of nonlinear functions is like the study of non-elephants. What he meant was that there's an infinite number of ways to be nonlinear. It's within that vast space of possibilities that complex systems exist; there are all these things that we have to study.

Last, complex systems product unbelievable novelty; all these interacting diverse things can produce all sorts of new, exciting things. In the market we have green ketchup, pancakes in a can, we have heated socks, in fact; I think that stuff is just awesome, it's totally awesome. As cool as these products are, though, they sort of pale in comparison to what evolution can do. Let me give two examples as we sort of conclude here. The first one's taken from George Williams in his book *The Pony Fish's Glow*. Suppose on a snowy day you're looking up into the sky and you see snow coming down. If you look at it, it actually appears a little bit dark, and that's because the snow is lit from up top. So if you wanted to make the snow invisible, what you'd want to do is you'd want to put little lights on the bottom of the snowflakes and that way you wouldn't see them.

So now suppose you're a fish designer—you're designing a fish—and you want that fish to be invisible from below; you don't want predators to be able see it. You'd want the belly of that fish to be white, but you'd want to have a light in the belly of the fish as well so that it doesn't appear dark from below. It turns out ponyfish have a light in their belly. No really, they have phosphorescent tummies—that's not a technical term—and what's really impressive, as they go deeper in the ocean, as the ocean pressure increases, the amount of the light changes so that the glow varies so that they become less visible as the ocean gets darker. The light in their tummy adjusts so they can't be seen.

Here's one other one; it's one of my favorites. Sea squirts when they're in their larval stage have the power of locomotion, so what they do is they move around and they find a good little rock to set in. Once they get in that little piece of coral or rock, they adhere to it and they spend the rest of their lives filtering water in pure tranquility. But then, what do I do with this brain? I have this brain and I was using it for locomotion, and now I don't need it anymore and it has all this protein in it. So they eat it; that's what they do, they just eat their brain. They sort of move around, they find their place in the world, and then they eat their brain. Incredible; just incredible.

So now we have it all in a nutshell: Complex systems are amazingly robust, and they owe this robustness to their adaptable parts. Yet, paradoxically, because they have these adaptable parts, they can collapse or they can have these huge events. Not all large events are bad—the Harry Potter books were a large event, and that's wonderful—but many of them are not: wars, epidemics, market collapses, and so on. Complex systems also produce emergent structures and phenomena with capabilities that far exceed those of the parts. Some of these phenomena—birds flocking, cultural formation, racial segregation, and goat paths—are self-organization; and those, if we study (and we'll see in this course), we can actually unpack how they work and see where the emergence comes from, even though there's no central planning, there's no CEO, and there's no middle management among the goats. Looking at things like consciousness we'll find that we can't quite figure it out, and that's why we're so interested in this.

So, on the one hand we have resilience and innovation; on the other hand we have overnight literary sensations and domino-like catastrophes. The complexity of the modern world is therefore not something to just sit back and passively accept, because we could suffer catastrophe after catastrophe. Nor is it something we should seek to control; to do so would be impossible. History is littered with failed attempts to control complex systems. Instead, we must seek understanding. Only through understanding do we have any hope of harnessing the enormous potential that the world possesses.

And yet it's true with intelligence and pluck we can harness complexity a little bit. Let me give one example as I close. Medical researchers now use recombinant DNA technology to produce proteins. These proteins are used in making what are called plasmids; plasmids are DNA molecules that have an extra chromosome that includes the genetic encoding for this relevant protein. These plasmids, medical researchers now inject into a cell line. When they do this, they're leveraging the complexity of the system—the cell's natural biochemical machinery—to translate that genetic information into multiple copies of the desired protein. The complexity of the biological process is harnessed to do good, and the result is lives that are saved.

If we can learn about complexity, if we can understand it, then eventually even if we can't control it, we can harness it. As Thomas Schelling once said—this is the guy who wrote *Micro Motives and Macro Behaviors*, Nobel Prize winner in economics—"If you're in the mood to be amazed" then complexity theory is going to be a great place to start.

Simple, Rugged, and Dancing Landscapes
Lecture 2

> **Dancing landscapes are complex. … Figuring out what to do in a complex situation isn't easy; and even if you do figure it out, what was a good idea today may not be a good idea tomorrow.**

In this lecture, we will use the idea of a landscape both as a metaphor and as a mathematical object. The simplest landscapes resemble Mount Fuji, which is shaped like a giant pyramid. Most of the landscapes that we will talk about will have many more peaks and valleys, much like the Appalachian Mountains. These are called rugged landscapes. People often conflate complex systems with rugged landscapes, but that is not quite right. The qualities of interdependence and adaptability in complex systems create landscapes that are not just rugged but dancing.

These three categories of landscape—simple (Mount Fuji), rugged, and dancing—are the main themes of this lecture. We will use these categories to understand why some problems can be solved optimally and some cannot. We will use the idea of a landscape to lay the foundation for how we will think about complexity. We will not just use the landscape as a metaphor. We will also have a formal, mathematical definition of a landscape.

Before we start, we need to define the two types of peak: local and global. A local peak is a place on the landscape from which a step in any direction is a step down in elevation. A global peak is the highest of all of the local peaks of a given landscape. Most of the time the global peak is unique. Mount Fuji landscapes are single peaked; the local and global peak are by definition one and the same. Rugged landscapes have many local peaks, and it sometimes can be difficult to find the global peak. Dancing landscapes can have a single peak or multiple peaks, but the key feature is that those peaks change over time.

In complex systems, agents adapt locally. If performance is considered as elevate on, then we can think of these agents as climbing hills. The local peaks are the best nearby options, whereas the global peaks are the best possible actions.

We will start by looking at the Mount Fuji landscape in the context of a real-world problem. The insight that experimentation can locate better solutions underpinned an approach to business that was pioneered by Fredrick Taylor in the early 1990s. It is often called Taylorism in his honor. Taylor solved a famous design problem involving the optimal size of a shovel. When charting the productivity of a shovel at increasing sizes on a topographical map, we find that it results in a two-dimensional Mount Fuji landscape. Since it is very easy to find the global peak in a Mount Fuji landscape, the productivity of an economy based on physical labor is easy to optimize by using scientific management.

Library of Congress, Prints and Photographs Division, LC-DIG-ppmsc-01808.

The world of Taylor is not the world of today. The world of today involves more rugged and dancing landscapes. Finding the highest point on a rugged landscape is not easy. The main reason is that the space of possibilities can be combinatorially huge.

At the turn of the 20th century, most problems were handled with manual labor on singular tasks, such as laying rails.

However, a large range of combinations is not the only thing necessary for a rugged landscape; the ingredients also have to interact. Once they do, the landscapes begin to have local peaks. These interactions occur within the choices of a single agent, as we can see when we consider the chain of effects created from the decision to remodel one aspect of a house. The simple rule is that the more interactions occur, the more rugged the landscape.

As difficult as rugged problems can be, they are not complex. To get complexity, we need to make the landscape dance. To explain the difference between a rugged landscape and a dancing landscape, we will look at two problems, one involving a milk distributor and one involving an airline. The problem of the milk distributor is difficult but remains fixed. The problem of the airline, however, contains multiple and interdependent actors; therefore, it dances. The difference is subtle. Interactions between our own choices are what make a landscape rugged. Interdependencies between our actions and the actions of others are what make a landscape dance. Interdependencies only come into play if the actors adapt. Hence complexity requires both interdependence

Let's do a quick summary of what we have learned about landscapes so far. We can think of the value of a potential solution to a problem as its elevation, so we can therefore think of a problem as creating a landscape. When a problem is simple, with no interactions, we get a Mount Fuji landscape. When our choices interact, we have a rugged landscape. If the elevations depend on the actions of others, and these actors exhibit interdependency and adaptation, then we have a dancing landscape.

In complex systems, agents adapt locally.

Why do all of these landscapes matter? Seeing the distinction between a Mount Fuji landscape and a rugged landscape helps us to understand why evolutionary and creative systems sometimes can find an optimal solution and sometimes cannot. The distinction between fixed and dancing landscapes has implications for how we allocate resources. Rugged landscapes have a good chance of repaying investment, whereas dancing landscapes do not.

The final and most intriguing insight that comes from these landscape models relates to what is called a standpoint, or perspective. In creative systems, there is no landscape. The landscape is determined by the way a problem is encoded. If a problem were encoded according to certain attributes, then the landscape might be simple. If the same problem were encoded according to other attributes, it might be rugged. The same logic holds for complex situations—how we encode them influences how quickly and effectively we can adapt. ■

Suggested Reading

Holland, *Adaptation in Natural and Artificial Systems*.

Mitchell, *An Introduction to Genetic Algorithms*.

Page, *The Difference*.

Questions to Consider

1. How have changes in technology made the landscapes facing firms, organizations, and governments more rugged?

2. Why might we commit fewer resources to formulating a policy on a dancing landscape then we would for a fixed, rugged landscape?

Simple, Rugged, and Dancing Landscapes
Lecture 2—Transcript

In this lecture, we're going to spend most of our time talking about landscapes. We're going to use the idea of a landscape both as a metaphor—we're going to think of ourselves as climbing up a landscape—and as a mathematical object, where we map the value of a function at a particular point as an elevation on a landscape; so elevation is going to play the role of function value.

I'm going to start with the simplest landscape imaginable: Mount Fuji. If you've never seen Mount Fuji, it looks like a giant pyramid or a cone, and it's relatively easy just by looking to find the highest point on Mount Fuji. This lecture, we're going to talk about how some real world problems from business and from ecology map into Mount Fuji. I mean that literally: We're going to literally map real word problems into Mount Fuji. Most of the landscapes we're going to talk about aren't going to be Mount Fuji; they're going to have more peaks and valleys, they're going to look a little bit more like the Appalachian Mountains. These problems are going to be harder to solve than Mount Fuji's, because just like it's much harder to find the highest point in the Appalachians, there are lots of peaks, there's lots of places where you could get stuck. So if you're standing at some point on the Appalachian Trail, you might see lots of paths pointing upward, and it's just not clear which description you'd take.

The Appalachians are what we're going to call a rugged landscape. In this lecture, we're going to learn how landscapes become rugged; and the answer's going to be that there are interactions between the variables or choices that an actor takes. For example, suppose you're designing a car. If you add weight to the car to make it safer you might think it makes a better car, but when you do that the car may be worse because now it needs a bigger engine to carry all that weight. So changing one variable to make things better might actually make the overall car worse.

People often conflate complex systems with rugged landscapes; they'll say a rugged landscape, a hard problem, is what complexity is. But that's not quite right, so let's think back to our opening lecture. Complex systems involve interdependent, adapting entities. What this means is the value of a solution—the height of a point on the landscape—depends on what other people do. In landscape terms, this means that you could be standing at some elevation, and then some other actor takes a move and the landscape literally dances (it goes up or down).

In this lecture, we're going to highlight how for complexity it's this combination of interdependence between the entities—people, firms, governments, species—and adaptation and moving around that cause the landscape to go up and down. Throughout, we're going to assume that the entities are connected in some way and we're also going to assume that these entities are diverse. But to get our bearings, we're going to keep connectedness and diversity at stage left for this lecture, because we just want to get a core understanding of how these landscapes get rugged, and also how they dance. In later lectures, we're going to get back to connectedness and diversity; we'll get them back into our story.

To get an idea of why dancing landscapes are so much more challenging than rugged landscapes—and again, dancing landscapes are what complexity really is—let's imagine for a second that you're sitting atop a peak in the Appalachian mountains and there's a rainstorm and you're watching the floodwaters rise. You're going to feel incredibly safe and secure. Even though the landscape is rugged, you're going to be at some local peak; you're going to feel fine. Now, suppose an earthquake happens, and there's a whole bunch of aftershocks, so the landscape starts buckle and shift. These tectonic events are going to raise and lower the ground under your feet, literally; so one minute you could be on a summit, the next minute you could be about to be submerged by rising rainwater. Staying dry, staying in a good solution when the landscape dances, requires constant awareness and frequent movement.

So these three categories—Mount Fuji landscapes, rugged landscapes like the Appalachians, and dancing landscapes—are the main themes of this lecture. Okay fine, you might say; but landscapes, I get that. But how does this have any relevance? Isn't this just all metaphor? First, as for relevance: We're going to use these three categories to help us understand why some problems can be solved optimally, both by humans and by evolution, and why others can't. We're also going to see why hard problems—like building an atomic bomb or curing cancer—are not the same thing as complex systems, like markets, ecosystems, ecologies, and international relations. We're going to see why the best approach to solving a difficult problem, or a rugged landscape problem, differs from the best approach to thriving in a complex world. Most important in all this is we're going to see how this landscape metaphor—this idea of a landscape—lays the foundation for how we can think about complexity. It's not just going to be as metaphor, though; remember we've got this precise mathematical definition of a landscape. The point in the landscape is basically your action, the solution; and the elevation is the value of that point.

Before we start, we need two definitions; these are intuitive but they're very important. We all know when we look at a mountain what a peak or summit is on a landscape; it's just the top of a hill. We want to make a distinction between a local peak—that's just a place on a landscape that if we step to the north, south, east, or west we'd go down—and a global peak. A global peak is going to be the highest of all the local peaks in the landscape. So Mount Fuji has just one local peak and it's also the global peak; the Appalachians have lots and lots of local peaks. Again, Mount Fuji landscapes are single peaked and rugged landscapes have lots and lots of peaks.

What we want to do is think about how you find a peak on a rugged landscape. If you had lots of time to walk around, of if you could traverse the entire range, then it would be easy for you to find the global peak. But that's typically not going to be the case; so you may get stuck on a local peak. When we get to a dancing landscape we've got a little bit more of a problem. It could be that they have only a single peak like a Mount Fuji; or it could be they have multiple peaks like the Rockies; but the key feature of these dancing landscapes is going to be that they move; they change over time. So if you wanted to stay on the global peak, you're going to have to keep moving; you're going to have to adapt and learn.

The idea of a landscape plays a central role in this course; we're going to keep coming back to it throughout. Here's why: In a complex system, agents are constantly adapting locally. These adaptations are attempts to improve performance or payoff. Performance or payoff we can think of as elevation; we're capturing that as elevation. What we can do is we can think of these agents in a complex system—firms, people, species—as sort of climbing on landscapes; trying to get to local peaks. And if they're lucky, maybe they get to global peaks. But that's not going to necessarily be easy to do. Now that we've got this idea—the elevation on a landscape represents how good our choice or action is; the higher the elevation, the better the solution—we can move beyond metaphor and we can start doing some real science.

First up: Mount Fuji. One of the points I raised in the opening lecture was that the world has become more complex. The drivers of that complexity have been increased diversity, more connections, greater interdependence, and quicker reaction times (you order the milk and they phone the cow). In the past, the core economic, social, and political problems just weren't as complex. That's not to say that complexity didn't exist, it's just that we sort of had this amping up—ramping up, if you will—of complexity. Let's go back to following World War II. A lot of the problems that we faced at that time were engineering problems: how do we build houses, cars, washing machines, etc. If we think back a little bit further to the turn of the 20th century, we see that many of the problems involved manual labor on singular tasks: laying railroad track, shoveling coal, and applying rivets. For these types of problems, a process of trial and error combined with good time management techniques was able to locate optimal solutions. In fact, the insight that experimentation can locate better solutions underpinned an entire approach to doing business that was pioneered in the early 1900s that has come to be called Scientific Management or Taylorism, in honor of Fredrick Taylor, who was its first great advocate. Henry Ford was one of the people who bought most heavily into Taylorism, and he set in motion something that's been called the efficiency movement.

What did Taylor do exactly? I want to take a famous example of his, and I want to show how this example translates into a Mount Fuji problem. Was this the problem the design of an engine, or even a light bulb? Nope, simpler than that: it was the shovel. I'm not kidding; Taylor determined the optimal size of a shovel. So how does he do this?

Suppose we start with a shovel with no scoop (so the technical word for this is a stick). If we're trying to shovel coal with a stick, we're going to get zero productivity; so the landscape's going to be very low. Nothing lifted, nothing gained. As we start to make the shovel larger, we can lift more and more coal. The way we're going to measure productivity of the shovel is how much coal an average worker can lift in an eight hour shift. As we increase the size of the shovel, productivity increases; so we can think of our landscape as getting taller. But at some point the shovel becomes too heavy, and productivity's going to fall off. Any further increases in the weight of the shovel are going to lower productivity even more. For example, if we were to make the shovel the size of a kitchen table or this stage, productivity would be down to zero.

What we can do is turn this real world problem into a mathematical function, a graph; and we can think of that graph as a landscape. What I can do is I can plot the weight of the shovel on the X-axis, the horizontal axis; and I can plot the total amount of coal that someone could lift in an eight hour shift on the Y-axis, the vertical axis. What we see—which we just explained—is that productivity's first going to increase as the weight of the shovel increases, but then it's going to decrease. If we think of this graph as a topographic map, what do we have? We have a two dimensional Mount Fuji. The slope goes up, reaches a peak, and the slope goes down. A problem like this is easy to solve—a Mount Fuji problem—whether we're doing math or whether we're just on foot, because in the case of the shovel, all we have to do is keep making the shovel bigger until it gets too heavy and then we stop; and we're done, we found the peak. By the way, in case you're interested, the optimal weight of a coal shovel is 21 pounds; and it turns out this weight is often regardless of what you shovel. This is why snow shovels are bigger than coal shovels. I've never seen one, but I'm guessing if you went to some factory where they use those Styrofoam peanuts, the shovels are probably just enormous; they're huge.

I want to think about the implications of Taylorism in an economy composed of physical laborers. This is the world of 60 or 70 years ago where people apply rivets, pound nails, attach doors, and make pies and cakes. In that sort of world, there are a whole bunch of Mount Fuji problems. What that means it scientific management is going to work, and it's going to work well, because it's going to find the global peak on problem after problem after problem. I'm going to just put aside Marxist critiques of this. Scientific management, this sort of repetitive labor, is going to be dehumanizing; it's not going to be a good way to spend your life, applying rivet after rivet. But that's not relevant to our discussion. What's relevant to our discussion is that the world of today is not the world of Taylor. The world of today involves rugged and increasingly dancing landscapes. That's what we want to go to next; we want to build up to dancing landscapes. But first, we have to talk about rugged landscapes.

So let's recall our mental representation of a rugged landscape: It's a mountain range—like the Appalachians, the Rockies, or the Himalayas—that have lots of peaks. Finding the highest point on a rugged landscape isn't easy. The main reason for this is the space of possibilities in huge; it's what we call combinatorially large. This size wouldn't be a problem if we had a topographic map, if we could just pull out the map and say, "Look, here's the highest point." But we don't. Not only that, we typically can't take leaps anywhere in the space of possibilities; we often are forced to search locally. Leaps, as we'll talk about later in this lecture, often require genius. Evolution is particularly constrained in its movements in this regard; it can't just make these leaps across space; it can't cross tigers with elephants to make "tigephants," which would be some sort of enormous grey beast with black stripes and sharp teeth.

When I say that these things are combinatorially huge—these problems; we can't make these leaps—what I mean is that there is an enormous number of possible solutions; an enormous number of things we can check. Let's say we have someone come over for dinner, and you say, "What would you like?" and they say, "Let's have chicken." You decide, "I want to make the best chicken dish in the whole world. How many choices are there? First, you have to pick the type of chicken: You could have free range, corn fed, grass fed. Then you have to decide what part of the chicken: Is it going to be the leg, breast, back, thigh? Maybe you cook the whole bird. Now you have to decide how you're going to cook it: Are you going to bake it, fry it, boil it, grill it, or slow cook it? Or maybe you just nuke the thing. Once you decide what type and how you're going to cook it, you have to decide sauce; are you going to have a rub? Let's say a rub; make it simple. Now you open your spice drawer and you see 50 spices. If you choose 3 of these 50, that gives you 20,000 spice combinations. 20,000! Now we have to decide which vegetable or grain are we going to serve as an accompaniment? You have hundreds of choices and literally millions of combinations of choices. So let's multiply all these things together—number of ways to cook it, number of spices, number of combinations—and what you end up getting is literally billions and billions of possible chicken combinations.

All these different combinations translate into lots of places to explore; but having lots of combinations isn't the same thing as being rugged. It just means that it's a big landscape; we still don't have that it's a rugged landscape. To get ruggedness, what we need are interactions; these interactions are going to make the landscape rugged. These interactions occur within the choices of a single actor. To see how they create peaks, we need to dig a little bit more deeply into an example; and I want to move from cooking the chicken into the idea of remodeling a house, because I think this will hit home with people a little bit more (no pun intended).

Suppose I have a house that has a small entryway, and I decide I'm going to expand this entryway; it seems like a good idea. But when I expand the entryway, what happens is I have to wipe out this bedroom closet. So what I thought was a good move—a move up on the landscape—has made the house worse, because now there's no closet in the bedroom; so turns out maybe the original design was a local peak and I just stepped off it. But let's not give up, we can keep moving forward. Let's just add another closet to the bedroom. That seems great. But turns out if I do this, now I have to wall over a window, so now the bedroom's too dark; so again, attempts to move up in the landscape have made me worse off. We're slipping in elevation; and the reason we're slipping in elevation is these interaction between variables. You could say, "Wait; let's just add a window to the bedroom." If we do that we have a better entryway, we have a closet, and we have a well-lit bedroom; everything's right with the world. Let's ignore the fact that we spent $10,000 making this better, let that go; the thing is we had to make a whole bunch of changes to get up one local peak to get to another one. To go up, we had to go down.

It's because these choices interact that getting to a higher peak requires multiple changes (entryway, closet, window). Here's the simple rule: the more interactions, the more rugged the landscape. The problem of the optimal size of the shovel isn't connected to anything; there are no interactions except what's being shoveled, so that problem's simple. Decisions in the home remodel are all interconnected—they're all interacting—and they're connect to various systems (electrical, plumbing, etc.) so the landscape is rugged.

The problem of evolving species is a classic rugged landscape problem. Suppose we wanted to improve the human. We seem pretty good; but we have some problems. One thing we might want is a blowhole on the top of our head, because then we couldn't choke and we'd get fewer diseases. But how would we do that? We can't just drill a hole right through the center of our brain; that's not going to be so easy. That decision would interact with other parts of our physiology.

As difficult as rugged landscape problems can be, they are not complex. They're not complex they're fixed; they stay the same. Ruggedness is often a component of complexity, but to get complexity, what we need is the landscapes to dance. To explain the difference between a rugged landscape and a dancing landscape, I'm going to describe two problems that seem very similar. The first is going to involve a milk distributer; and the second, an airline.

First, the milk distributer: Each day, cows produce milk, farmers harvest the milk, the distributer goes and picks it up and pasteurizes it, and then has to deliver it to different stores. Assuming the demand at these stores stays about the same and the cows produce about the same amount of milk, the distributer just has two problems to solve: first, which trucks to send to which farms; and then once it's pasteurized the milk, which trucks to send to which stores. It wants to do this in such a way to maximize profits. This isn't an easy problem; it's a rugged landscape, in fact. Why? Because shifting one farm from one route to another route may lower the costs to reassigning another farm—a neighboring farm—to that route as well. Finding the optimal solution to this problem is really difficult, and computer scientists spend a lot of time and use a lot of computational power trying to solve them. What's relevant for us, though, is the problem—as hard as it may be—stays fixed. The farms aren't changing location, the cows aren't moving around; so it's the same rugged landscape day after day after day.

Let's consider an airline, though. That's a similar problem: It has to decide where and when to route its planes; and it has to do this in such a way to maximize profits. This profit landscape is also rugged; shifting one flight may require changing other flights. What differs, though, is the airline's landscape dances. It dances because when other actors change their behavior, the airline's profits—and hence its landscape—changes, too. There are interdependencies between the actors.

Who are these other actors? First, there are the consumers. Adding a route—say Chicago to Hamburg, Germany—may increase demand on a whole host of flights to Hamburg, and then by necessity to Chicago. This doesn't happen for the milk; when the milk distributor adds a new stop—say a new farm—milk from the other farms doesn't line up and say, "Oh, wait, we want to get on the truck so we can go to this new farm."

Second, the other airlines also affect the landscape. When they shift or add a route, they alter our airline's profits. Let's suppose you're running an airline. What you do is you decide, "We have to solve this problem. We have to figure out how to route our planes." So let's hire some really bright statisticians, mathematicians, computer scientists, maybe even economists. We'll put these brilliant minds to work for months, and they're going to figure out the perfect set of routes and prices that give us the maximum profits. They figure all this out and we decide that on June 10, we're going to implement these changes. You wake up on June 9 and you see that a competing airline has implemented a whole new set of routes and prices; so they've completely changed where they're routing their plans. It turns out it, too, had hired mathematicians, statisticians, and computer scientists, but it moved first.

What does that mean? It means that the problem we just spent—the rugged landscape that we just spent—spent six month trying to solve is gone. Poof; it's just gone. The landscape has danced; there's been a tectonic shift. Mountains have gone up and down; hills rise where there were once valleys; peaks sink to the floor. It was once a good solution; but may not be a good solution.

Let's go back to our house example: It's as though you've spend months formulating an ideal renovation of your house only to wake up on the morning construction's going to begin and find someone's replaced your house with an entirely different house. Once there was a hallway; now there's now a den. Once where your kitchen was, there's now a bathroom. And by the way, you now have a screen porch. That's what I mean when I say life on a dancing, rugged landscape is complex.

Here's the subtlety, and it's a really important one: What makes a landscape rugged are interactions between our own choices; we saw that in the case of designing the house. What makes a landscape dance, and a situation complex, are interdependencies between our actions and the actions of others. When one player's payoff—the height of their landscape—depends on the actions of another one, that's an interdependency. Those interdependencies only come into play when other actors are adapting; if they move. So complexity requires both interdependence—their actions affect ours—and adaptation. Note also that just as increasing the number of interactions increases ruggedness, increasing the number of interdependencies increases complexity; though only up to a point, as we're going to see in the next lecture.

Your world—your daily life—would be much easier if landscapes would stay fixed, but they don't. Because they don't, it's not necessarily the case that people, governments, firms, organizations, or even species are always doing the optimal thing. Sometimes they're doing the thing that used to be optimal, but no longer is. To borrow a phrase from Charles Lindblom, a political scientist, people are muddling through.

Quick summary: We've thrown out a bunch of ideas here, and I just want to get our bearings. We can think of the potential solution to a problem as an elevation on a landscape, and if we numbered those solutions—let's say like from one to "n"; we plotted their values—in some cases we get a Mount Fuji. That's going to be an easy problem; we just climb to the top and hold up our 21-pound shovel. But when our choices interact—which is the case when designing a house or routing the milk distributor—we get a rugged landscape; we get something that looks like the Appalachians or the Rockies. Finally, if our payoffs depend not only on our own actions but the actions of other people, we get a dancing landscape. A dancing landscape requires these interdependencies and adaptations. To keep this straight: interactions between an individual make the landscape rugged; interdependencies between people make the landscape dance.

Then you might ask who cares? Why does all this stuff matter? First, seeing the distinction between a Mount Fuji landscape and a rugged landscape helps us understand why sometimes evolution or human systems can locate optimal solutions—the 21-pound shovel—and why sometimes they don't. Suppose you're out in a field and you look at the face of a sunflower; you're going to see a tight spiral of seeds. As the sunflower grows, these seeds sort of attach themselves sequentially starting from the center and spiraling outward. You can think metaphorically of a tractor just driving out from the center of the pod head and dropping seeds at some distance. We can measure these at an angle, because the tractor is driving in circles. That angle in a sunflower is approximately 137.5 degrees; so think of the hands of a clock at 10:15.

Suppose you ask yourself the question, "Suppose I want to pack as many seeds as I could into the head of a sunflower? I'm going to begin at the center of the seed and sort of spiral out; what should that angle be? How often should I drop a seed?" The answer turns out to be 137.5 degrees. That's sort of amazing; let's see why that's true. Suppose the angle were 90 degrees; so if every 90 degrees you dropped a seed, then what you'd get is four straight lines starting off from the center of the flower. You'd drop one at noon, you'd drop one a 3:00, you'd drop one at 6:00, and you'd drop one at 9:00. Then you'd do that again: one at noon, one at 3:00, one at 6:00, and one at 9:00. You want to avoid piling up seeds at the same angle, so the way to do that is to choose an angle that's an irrational number. The math here is very complicated, but the trick is to choose an angle proportional to something known as the golden mean, which is approximately 137.5 degrees. This is something that mathematicians figured out. But here's what's incredible: Evolution figured this out as well. How did evolution solve this?

Let's think about it. Early on, sunflowers didn't break out a chalkboard and prove some theorems about exactly what the golden mean is or what this angle should be; there was no Taylor in the world of sunflowers. What happened is sunflowers evolved with different angles for seed placement; so as the angle got closer to 137.5 degrees, they could pack in more seeds which meant more flowers. But once they went past 137.5 degrees, this led to fewer seeds and therefore fewer flowers. In other words, if you think of the angle as the horizontal axis, what we get in the packing seed problem is a Mount Fuji landscape; a single peak problem. It's not at all surprising that evolution could figure this thing out.

If we contrast the seed packing problem with the problem of designing a house, the house problem is more rugged. That's why architects keep coming up with new and better houses, because the landscape there is rugged. Designing a house requires lots of choices and those choices interact. Hilary conquered Everest, that's true; but the perfect house has yet to be built (Fallingwater by Frank Lloyd Wright notwithstanding).

Second, the distinction between fixed and dancing landscapes has massive implications for how we allocate resources. Building an atomic bomb, as hard as that was, was a rugged landscape problem; the landscape didn't start to dance until after the bomb got dropped. Regardless of who struts and frets upon the stage, a bomb is a bomb is a bomb; therefore, the government could solve this problem by doing what? By squirreling away brilliant physicists in the mountains of New Mexico and letting them just go at it; the landscape was going to stay fixed. Yes it was rugged, but it was not going to dance.

If you have a rugged landscape problem like building a bomb, it makes a lot of sense to devote serious resources for a long time: let's cure cancer; let's build a more efficient and safer nuclear power plant; those are rugged landscape problems. But once the landscape starts to dance, it makes a lot less sense to devote too many resources or spend too much time. Why? Because time is of the essence; as soon as you start trying to solve the problem, it may move. This is one reason why government panels to reform social security or to overhaul the tax code come up with recommendations in a few months rather than a few years. If problems change quickly, so must solutions.

In the case of the Manhattan project—building the bomb—if the landscape had been dancing, the physicists could not have been tucked away. They'd have been better situated in, well, Manhattan, right in the seat of action. That's true for most modern knowledge workers; we're all toiling on a dancing landscape. The closer a person is to the action, the more she's aware of the changes in the landscape. That's just one of many reasons workers in larger cities are more productive on average than workers in rural areas: they're more aware of how the landscape is moving.

A final and most intriguing insight that comes from this landscape idea relates to what philosophers call standpoint and what we commonly call perspectives. What I'm going to say next is going to sound a lot like a phrase from some sort of New Age, deconstructionist, Zen Buddhist; and it's an idea I introduced in a book I wrote a couple years ago. The idea is this: In creative systems—not evolutionary systems, but creative systems—there is no landscape. Let me say that again with sufficient depth and gravity: In creative systems, there is no landscape.

What I mean by that statement is the following: The ruggedness of a landscape depends on how it's encoded. If I encode a problem in one way, it might be Mount Fuji. If I encode it another way, it might be rugged. It depends on how we represent it mathematically. Let's go back to designing a shovel. If I take 20 shovels and arrange them by weight, we know we get a Mount Fuji; we saw that from Taylorism. But suppose I borrowed these 20 shovels from neighbors. In that case, instead of encoding them by weight or size, I might encode them by the name of the person who gave me the shovel. Is that nuts? Yes; but I could do it. So the first shovel I got from Mr. Assan may be better than the second shovel from Mr. Barber; if that's true, there's a peak at Mr. Assan. Mr. Campbell's shovel, which is third, might be better than Mr. Barber's and might be better than the one I borrowed from Ms. DeVerne; so there'd be another local peak at Mr. Campbell's. So if I plot the entire landscape of shovels arranged by name, I'm going to get lots of peaks. It's going to be a rugged landscape.

Many of the greatest scientists of all time—Newton, Einstein, Mendeleyev, Curie—made breakthroughs by developing new ways for seeing the world; new encodings. The periodic table is a new encoding; the space-time continuum is a new encoding. These great minds made breakthroughs because they were climbing different landscapes then the rest of us. This allowed them to basically take what were rugged landscapes and turn them into Mount Fuji landscapes.

Let me sum up: We've seen how we have Mount Fuji landscapes, rugged landscapes, and dancing landscapes. Dancing landscapes are complex. That complexity results from interdependencies with other adapting entities. Figuring out what to do in a complex situation isn't easy; and even if you do figure it out, what was a good idea today may not be a good idea tomorrow. The lives that we lead are an endless sequence of adaptations on a dancing landscape. And so it is for the people with whom we collaborate and compete; it's the world in which we live. And that world is, in a word, complex.

The Interesting In-Between
Lecture 3

To dig more deeply into how the attributes of interdependence, connectedness, diversity, and adaptation and learning generate complexity, we can imagine that each of these attributes is a dial that can be turned from 0 (lowest) to 10 (highest).

We are going to spend this lecture twisting those dials to see which combinations lead to complexity and which do not. Contrary to what we might expect, we will not get complexity at the extremes. Complexity exists in a region that I like to call the interesting in-between. First, we need to come to a common understanding of what behaviors a system might take on. For that, we need to categorize both the system and its initial states. The behavior of a system can depend on its state and the rules followed by its parts.

Physicist Stephen Wolfram divides the behaviors of systems into four classes. Class 1 behaviors are stable, single-point equilibria, like a ball at rest at the bottom of a bowl. This behavior is said to be resistant to perturbation. Class 2 behaviors are called periodic orbits. A periodic orbit is a regular sequence of states, like the cycle of a stoplight. Class 3 behaviors are chaotic, meaning that they are extremely sensitive to initial conditions. An example of this would be the proverbial butterfly that flaps its wings and creates a hurricane. Class 4 behaviors are complex. They have regular structure but they also have high information content, meaning that they would take a long time to describe. What has to be true about a system for it to fall into one of these four classes? Experiments with our dials will show us.

Let's start with the interdependency dial. With the interdependency dial set at 0, each person does what he or she wants to without any concern about what others do. Think of a person choosing to wear a sweater when it is cold. If we ramp up the interdependency to a moderate level, as in a teenager trying to decide on a cool sweater to wear, we get complexity. If we turn the interdependency dial up higher, so that the teenager is trying to gauge the coolness not just of the sweater but of every item of clothing, we get a chaotic mess. One student changing his initial outfit could result in a drastically different path of outfits for all students. It will not always be the case that high interdependency leads to chaos. Sometimes it just causes an incomprehensible mangle.

The next dial is connectedness. Whereas interdependence refers to whether other entities influence actions, connectedness refers to how many people a person is connected to. If a person is completely disconnected from everyone else, no one else can have any effect on that person's actions. The result is not complex. If we hold the interdependency at a moderate level and raise the level of connectedness, we come up with some interesting results. We can see this in the greeting game, in which people "best respond," meaning that they respond in the way they remember their connections responding before. At a somewhat low level of connectedness, equilibrium is established rather quickly. At a moderate level of connectedness, it can take a long while for equilibrium to be achieved. At a high level of connectedness, equilibrium is once again achieved quickly.

Almost all games studied in game theory have either two players or infinite players. As a result, game theory tends to ignore the interesting in-between, where complexity happens.

The same phenomenon exists if the game is made more complicated, as in rock-paper-scissors. Although in this game the best-response technique is used to mismatch others instead of match them, complexity is still only found in situations of moderate connectedness. If we apply the same experiment to the *Escherichia coli* found in our bodies, we find a real-life rock-paper-scissors game in which the states can be labeled sensitive, toxic, and resistant. Almost all games studied in game theory have either two players or infinite players. As a result, game theory tends to ignore the interesting in-between, where complexity happens.

We move on to diversity. When we say diversity, we mean differences in types. We do not mean variations. If we take chemical elements as our different types and we adjust them from no diversity to moderate diversity to high diversity, we find the same pattern as in interdependence and connectedness: from simple to complex to a mess. The same thing holds for ecosystems, with the exception that a high diversity of species can sustain complexity if interactions with other species are restricted, as is the case with niches. In most cases, however, high diversity leads to the collapse of the ecosystem.

We are now ready to study the relationship between complexity and adaptation and learning. We should think of this dial not as increasing the speed with which entities adapt but as increasing the intelligence of these entities. If there is no adaptation or learning, simple parts follow fixed rules and remain in equilibrium. A little learning or adaptation goes a long way. It allows the parts to figure out how to interact with one another to create a coherent, complex whole. When we turn the learning dial all the way up, we often get equilibrium again. If everyone is optimizing relative to what everyone else is doing, there is no need for the system to change, and we get what is called in game theory a Nash equilibrium.

When we say a system is complex, we mean that it produces interesting nonperiodic patterns and emergent structures and functionalities. We have found that this complex state tends to lie in a region of moderate interdependence, moderate connectedness, some diversity, and some adaptation. Implicit in what we have learned is an explanation for why we see so much complexity in the social world. As social beings, we are connected, interdependent, diverse, and smart—but all in moderation. ■

Suggested Reading

West, *Scaling Laws in Biology and Other Complex Systems*.

Questions to Consider

1. In a complex system, does increasing the number of connections have the same effect as increasing interdependencies?

2. If government becomes an incomprehensible mangle, how might it respond to regain vibrancy? Alternatively, if the same government became predictable and ineffectual, how might it achieve vibrancy?

The Interesting In-Between
Lecture 3—Transcript

In the first lecture, we discussed what complexity was, its attributes, and why it was important. In the second lecture, we introduced the concepts of Mount Fuji landscapes, rugged landscapes, and dancing landscapes; and we made a key distinction between difficult problems—which is finding a peak on a rugged landscape—and surviving in complex systems; these are situations in which the landscape dances and we need to adapt and learn in order to survive and be successful.

In the second lecture, we observed that interactions between the multi-dimensional choices of the same entity produce a rugged landscape, and that it's interdependencies between the actions of different entities—different people, different firms—combined with adaptations that are going to make the landscape dance; that create complexity. Implicit in that characterization of interdependence was the idea that the entities were connected in some way. It was also the case that in our many examples the entities were for the most part diverse.

What are we going to do in this lecture? In this lecture, we're going to dig more deeply into the four main attributes of a complex system—interdependence, connectedness, diversity, and adaptation and learning—and we're going to see how those things generate complexity. What we're going to do is we're going to get dialed in to complexity, and I mean that in a literal sense. I want you to imagine yourself sitting in front of a box. On that box are four dials or knobs, and above one of the dials it says diversity; above another it says connectedness; above another it says interdependence; and above another it says adaptation or learning. Each of these dials can be set from some number between 0 and 10. If we set the diversity knob at 0, that means every entity is the same; every agent is the same. If we crank it up to 10, they're all different. Same goes for connectedness: if we set it to 0, nobody's connected to anybody; if we set it to 10, everybody's connected to everybody else. The same holds for interdependence and adaptation: each of those knobs we can set at 0, or we can set them all the way up to 10. For example, if adaptation is equal to 0, that means no one's changing their behavior; the landscapes are fixed. If it's at 10, everybody's instantaneously jumping to a peak.

That's why I said we're going to get dialed in to complexity, because we're going to spend this lecture twisting those dials to see what combinations lead to complexity and what don't. What's really fun about these experiments is they're not passive, observational lab experiments; we're not going to put on white lab coats and sit back and watch what happens and record it in our log book—"the chemical reaction appears to be emitting a toxic odor"—these are going to be active thought experiments. So when we twist the dials, that's just the beginning; once we set the values, we have to sit back and think through what's going to happen.

As we turn the dials, we're going to find that contrary to what might have expected, we're not going to get complexity at the extremes; we're going to find that complexity exists in a region that I like to call "the interesting in between." What do I mean by that? I mean that if things are too connected—if everything connects to everything else—we tend not to get complexity; instead, we're going to get statistical regularity. On the other hand, if connections are rare, then a system just doesn't have enough going on in to produce complexity. Therefore, for complexity to emerge, it must happen in this interesting in between space; the space between boredom of no connections and the unproductive mangle of too many connections.

We're going to see how this insight of complexity lying in the in between region plays out for all four attributes; all four dials. But before we can do that, we need to come to a common understanding of what sort of behaviors a system might take so we can identify situations that are complex from situations that are not. We're going to use a classification that Stephen Wolfram, a physicist, created. Before we start, though, I've got to be careful in explaining what it is we're going to be categorizing. We're going to be categorizing the system itself, the rules it follows, and the initial states. What do I mean by that?

Let's take a family. A family consists of interdependent, diverse actors (it's true of my family anyway). If we sit that family down at the dinner table, we might find that they settle into a nice stable pattern of behavior; so we might say the family's stable. However, let's take the same family and put them in a car for 36 hours. In this case, their behavior is going to be, well, let's just say complex: We might have long periods of silence, followed by violent eruptions. We can't say that the family is stable, and we can't say the family is complex; it can be either one. Which one it is depends on the state the family starts out in, whether it starts out at the dinner table, or whether it starts out in the car. The same is going to be true of any complex system. Its behavior can depend both on its initial state and the rules followed by its parts. That's how we're going to think of systems.

Once we think of systems this way, Wolfram breaks them into four classes. Class 1 behaviors are stable; these are what we call single point equilibria. Remember the case of our ball sitting at the bottom of the basin, that's a stable equilibrium; if we push the ball up a side, it comes back. Now you might say a ball in a bowl isn't very complex, and that's not; but it's also true that we can have systems of diverse interacting agents that end up looking like the ball in the bowl. This would be true of a market: If you go to a farmer's market, you might see a fairly stable set of prices and sales; so the market's in equilibrium even though it has diverse interacting agents.

The second class of behaviors are periodic orbits. A periodic orbit is a regular sequence of states; so a stoplight cycling through red, yellow, green is a periodic orbit. The earth rotating around the sun is also a periodic orbit. Periodic orbits don't need to be physical. If I write down a mathematical model of predators and prey, I'm going to get cycles in which populations of rabbits and foxes oscillate in a regular pattern.

The third class of behaviors we're going to consider are chaotic. By chaotic, I mean that they're extremely sensitive to initial conditions. If we take two initial states that differ just by a tiny bit, we're going to see that their resultant paths are going to diverge sufficiently that in a relatively short period of time that we may not be able to tell the two apart. Metaphorically, this means that a butterfly might flap its wings and redirect a monsoon or a hurricane six months later; that's what we mean by chaos.

The final class of behaviors, what Wolfram calls Class 4, are what we call complex behaviors. Like periodic orbits, complex behaviors have regular structure; but unlike simple periodic orbits, these patterns are longer and they have what mathematicians call high information content. Roughly speaking, this means it would just take a long time—a lot of information—to describe them. That's why life is complex; it's not easily explained. Stock markets are complex, and international relations, and families in the car. System behavior in these cases is not regular—no family, company, or country runs like clockwork—but they're not chaotic either. They lie in this "interesting in between"; they have structure, but the structure isn't easy to define.

So now I have this really interesting question: What has to be true about a system in order for it to fall into one of these four classes? In other words, what makes a system stable, and what makes a system complex? Let's reach into our box with the dials and do some experiments. Let's start with the interdependency dial set at zero; so each person is completely independent of anyone else, she just does what she wants without any concern about what others do. What might be an example of this? Suppose you think about what to wear: Do you put on a sweater? That decision has no impact on anyone else. If we looked at sweater-wearing behavior, or the color of people's sweaters, we could just count up what happens on average, and we're going to find that it's probably an equilibrium: the colder it is, the more people who wear sweaters; there's nothing complex.

But suppose we ramp up the interdependencies. Let's think about junior high school students and they're wearing sweaters. Now these students are thinking, "I want to wear sweaters that look like my friends' sweaters, but I don't want to wear sweaters that look like the people I don't like's sweaters." Once we get this, we have interdependencies, and we have the potential for complexity; and we end up with junior high school students standing in front of the mirror perplexed about what to wear. That is complexity; it's caused by interdependence.

Let's turn that dial even more, so now every student cares about what every other student wears, and every single item that the other students wear. Now we've drifted beyond complexity, and we just have a mess; because if one student changes his outfit, this could result in everybody else changing their outfits. We're going to get something that's closer to chaos; we're going to have extreme sensitivity to initial conditions: one change in an outfit leads to a mass set of changes. It's not always going to be the case that too much interdependency leads to chaos; sometimes we get—remember?—what I've called an incomprehensible mangle. If I let a 10 year old loose in the kitchen and just let them start mixing up ingredients, they're going to throw in all sorts of things but nothing interesting is going to emerge; it's just going to be pretty much a mess. So what we get—and this is the key point—no interdependency, it's not complex; too much interdependency it's either chaotic or a mangle; complexity happens when interdependency is in the middle.

Recall that interdependency refers to whether other entities influence actions, while connectedness refers to how many people a person connects to. First you can imagine someone disconnected from everyone else. To be disconnected means you have no interdependencies; no one has any effect, so you just do what you do and there's no complexity. So let's hold interdependencies at some sort of constant or moderate level and just change the connectedness. What I mean is that I'm not going to change how much someone matters you, but I am going to change how many people matter to you.

To put some structure on this, I'm going to consider two games. These are formal games that game theorists—people who study strategy—consider. The first one is called a pure coordination game. In the pure coordination game, the object is to take the same action as the people you're connected to. An example of a pure coordination game that we play every day is the greeting game. In the greeting game, when you meet someone you have to decide are we going to shake hands, are we going to kiss on the cheek, are we going to hug? You have to pick a simple rule and you have to follow it; and it's important that you do the same thing as your friends, because if you go to shake hands and they go to hug it's sort of awkward.

Let's supposed we have a network of people who are playing this game and they're trying to decide on a rule. What we're going to do is we're going to assume that they best respond; we're going to use this rule throughout the lecture. By that I mean they're going to take the best response—they're going to do the best thing—given what their friends did yesterday. So if yesterday your friends were mostly shaking hands, today you'll shake hands; if yesterday your friends were kissing, then you'll kiss. If the system is hardly connected at all—say each person has like one friend—then the system is going to go to an equilibrium, and some pairs of friends will shake hands, some will kiss, and some will hug.

If we make the system a little bit more connected—partially connected—initially, some people will shake, some people will kiss, some will hug, some of the shakers may switch to kissing, some of the kissers may switch to hugging, and so forth. After a while—perhaps after a long while—we can show that this system is going to settle down into equilibrium, but it's going to take a long time, and pretty much everyone will be doing the exact same thing.

Now let's turn the dial even further: Let's suppose that everyone is connected to everyone else. If everyone's connected to everyone else, people are going to very quickly converge on whatever action is most popular—kissing, shaking, or hugging—and the result will be an equilibrium that arises almost instantaneously. In this pure coordination game, what we get is if connections are low or if connections are high, the system just goes to equilibrium. But if connections are in this intermediate range, it could take a long, long time to get to that equilibrium; so we're going to get some complexity, and then an equilibrium.

Now I want to change the game—now that we have this idea about connections—to rock, paper, scissors. In rock, paper, scissors remember scissors cut paper, paper covers rock, and rock smashes scissors. Rock, paper, scissors differs from pure coordination in that now I don't want to match the people I'm connected to, I want to mismatch them; I want to do something different than they're doing. Let's suppose that we're very loosely connected—I just have one friend—and supposed that the person playing I'm playing is playing paper, then I want to play rock. Let's assume that we best respond; this means that the person who is playing paper which beat rock—paper beats rock—is going to stick with it, because it's the best thing to do against rock. But the person who is playing rock is going to say, "Wait a minute, my opponent's playing paper so I'm going to switch to scissors." That's what we call a best response, because the rock will smash the scissors.

In the next round, the scissors person is going to stand pat because he won—the scissors cut the paper—and the paper person is going to switch to rock. If we keep going with this logic, we'll see that we're going to get a cycle, and that cycle is going to be of length six. The first player's going to play paper, paper, rock, rock, scissors, scissors; and he'll do that again and again and again. The second player will play rock, scissors, scissors, paper, paper, rock. Each of them is going to win half the time and we're going to have what we call a periodic orbit; what Wolfram called a Class 2 behavior.

Let's turn the connectedness dial all the way up, so that everyone plays everyone else. Everyone in this whole population of players is playing every other person; so the first time these people play, let's suppose the majority of the people are playing rock. That means the next time everyone should choose paper, because paper's the best response. But then the next time, everyone should choose scissors, because scissors is the best response to paper. Then after everybody chooses scissors, the next time everyone should choose rock. Here we have a simple three cycle: it just goes rock, paper, scissors, rock, paper, scissors. What we see: not connected, we get a cycle; fully connected, we get a cycle.

What about in between? What if people are somewhat connected? In this case, it gets complex. To make it simple, let's put people on a checkerboard—I'll explain why in a second—and let's assume that each person plays his eight neighbors; if you're on a checkerboard there are eight squares around you. If you run this model with each square best responding, you're going to get very complex dynamics because everyone has slightly different neighbors who are playing slightly different strategies. You're not going to get simple cycles, you're going to get all sorts of interesting patterns that ebb and flow; you're going to see people sort of taking these weird patterns through rock, paper, scissors.

I recognize that this seems very, very abstract, but the reason I did this example is because it's very real. In our guts at this very moment, each one of us has several types of *E. coli*. Some of these *E. coli* are sensitive, others are resistant, and others are toxic. These *E. coli* are in effect playing a game inside our gut, and if we think of a head-to-head matchup, the sensitive *E. coli* are going to beat the resistant *E. coli* because the sensitive *E. coli* are simpler; they're simpler so they can reproduce faster, they have shorter DNA. However, the sensitive *E. coli* get beaten by the toxic *E. coli* because they can destroy them; the toxic *E. coli* get too toxic for the sensitive one. The resistant *E. coli*, because of their longer DNA, can defend against the toxic; so resistant defeats toxic. In short, in the lining of your gut you have a real-life game of rock, paper, scissors; but instead it's called sensitive, toxic, resistant.

There's an article published in the journal *Nature* by Marcus Feldman, Benjamin Kerr, Margaret Riley, and Brendan Bohannon. They did the following experiment: They took these three types of *E. coli*—they just scraped them out from someone's gut—and put them on a slide (the kind you'd put under a microscope). This is an analog of our checkerboard experiment—playing rock, paper, scissors—but it's real, it's with real *E. coli*. What did they find? They found complexity; what they found were these complex, diverse patterns of *E. coli*. To discern whether these patterns depended on connectedness, they then took these same *E. coli*, put them in a beaker, and they stirred them. By stirring these *E. coli*, that meant that every *E. coli* was connected to every other *E. coli*; so now we have full connectedness. What happened then? What happened is one of the *E. coli* won out. The system wasn't complex at all; you get an equilibrium. It's interesting; we see real-world science agreeing exactly with what we saw from our simple thought experiment. If we had a moderate level of connectedness, what we got was an interesting phenomenon; we got complex outcomes. But if we had too much connectedness, with every *E. coli* connected to every other *E. coli*, we got an equilibrium.

Here's where it gets even more interesting: If you take a game theory course (and my Ph.D. was actually in game theory) you study the strategic behavior of different players—these can be firms, these can be political entities, these can be two people playing tennis—it shouldn't shock you to find out that most of the results we have in game theory consider either two players, which are low connectedness, or an infinite number of players playing one another, which is full connectedness. With two players, it's not too hard to figure out what to do—you just figure out what the other player's going to do and you respond—and with an infinite number of players you're just responding to this population, the statistical regularity; and so again what you get is predictability. It's pretty easy to figure out what to do on average; so that's not very hard either.

What's interesting here is when you have moderate levels of connection, each person is playing sort of a different game in a sense; even though they're playing the same fundamental game, they're playing a different set of players and they confront a different situation, a different landscape if you will. The game becomes too complex to analyze using mathematics; so as a result, game theory tends to spend its time ignoring the "interesting in between" and concentrating at the dial being turned to low connectedness or high connectedness.

We're halfway home. Too little or too much interdependence, or too little or too much connectedness, and we don't get complexity. Complexity happens in the in between. Let's go to diversity now.

When we say diversity we need to clarify what we mean; diversity of what? If we're talking about an ecosystem, we might mean diversity of species. In an economy, we might mean diversity of firms or products. In a chemistry classroom, we might mean diversity of elements. What we don't mean is variation, which is slight differences in the members of a population; we're going to revisit this distinction in an upcoming lecture. Here, when I talk about diversity, I mean differences in types. Different types have different functionalities.

Let's begin in the chemistry classroom to see how levels of diversity influence complexity. Let's use elements as our types. If I have no diversity—if I just have a bunch of carbon atoms—not much interesting happens. But when I start to mix together a few elements, I might get some stable compounds. If I mix sodium and chloride, I get salt; if I mix hydrogen and oxygen I get water. The fact that salt and water differ in kind from their parts is emergence—something we talked about before and we'll talk about later— but for the moment what I want you to recognize is that if we mix a few elements together, we get interesting things popping out; like when we mix baking soda with vinegar and get a volcano. So it should be pretty clear here: no diversity, not much happening; increase the diversity so that entities with different functionalities interact and we can get something complex. But what if we go crazy and just start mixing together a whole slew of elements: carbon, copper, hydrogen, lithium, beryllium, etc.? Well, we're likely to get grey goo; just a mess. So again we see this same sequence of phenomena as we turn our dial: We go from simple, to complex, to a mess.

But wait a second, you might say; what about ecosystems? Does this relationship hold? Let's take a low diversity ecosystem like the Sahara Desert. The Sahara is bounded on one side by the Atlantic and on the other side by the Red Sea. Roughly speaking, it covers the top half of Africa; it's about the same size as the continental United States. It's mostly sand dunes and buttes; but it's not barren of animal life, you can find antelopes, gazelles, hyenas, there are even some badgers and gerbils. The best estimates are there are about 100 species of mammals and 100 species each of birds and reptiles in the central Sahara, and maybe 500 plant species. Given that the central Sahara gets between 5 and 25 millimeters—millimeters!—of rain a year, that's really not bad, but it's probably not complex.

By contrast, let's look at the Amazon: The Amazon River basin is estimated to have over 5 million species of flora and fauna. One 25 hectare region of the Yasuni Forest in Ecuador, which has been studied, supports over 500 species of birds and 1,100 species of trees. 25 hectares is about 50 football fields; it's about the size of the Mall of America. To put that number of trees (1,100) in perspective, the United States has approximately 850 species of trees. Due to deforestation, each week the Amazon is thought to lose nearly 1,000 species—that's more than the entire Sahara contains—that's in a week.

Despite all this diversity, the Amazon rainforest is not grey goo; it's complex. Here it seems like we've turned the dial a long way. Does that contradict the idea that complexity happens in between? No; here's why: The diverse Amazon ecosystem has been constructed over a long time with each species adapting and evolving to fill in a particular niche. This is a process called niche assembly, and it requires a restricting of interactions with other species. If rather than construct an ecosystem we just randomly toss together species and see what happens, we won't get complexity. We're more likely to see a collapse in the number of species. What I'm saying here is even though diversity's high, the system is responding by lowering interactions.

This isn't just intuition. Robert May constructed a theoretical model in which to explore how robustness of ecosystems correlated with diversity. In this model, he created species that ate and were eaten by other species, and he just selected them at random (who eats what). He found that as he increased diversity beyond a certain point, robustness decreased; the complexity went away and you ended up with systems that just collapsed. These systems were not likely to sustain complex dynamics.

How do we make sense of all this? Think back. May did something similar to what we did in our chemistry lab: He just randomly mixed stuff together. If we have random diverse types, then we get the result that too much diversity leads to either a mess or a collapse. However, in the Amazon rainforest, the species have had a long time to construct their behaviors and their niches so as to create a robust, complex whole. Therefore, it's possible for systems with increased diversity to be more complex, it's just not likely to happen by chance. So, if we think about cranking up the diversity dial, we should not expect that as we add diversity we get more complexity. We could get the opposite: We should get a collapse and a decrease in complexity. This is why ecologists worry so much about invasive species such as cats in the Galapagos, and plants like purple loosestrife or garlic mustard here in the United States. These are increases in diversity that do not increase complexity; instead, they destroy it.

We're now ready to study the relationship—the last one; the last dial— between complexity and adaptation and learning. If we think of turning this learning and adaptation dial, we don't want to just think of turning the rate at which these entities adapt, because then all that would do is make the system go faster and there's really not much interesting to talk about; instead what we want to do is we want to think of turning this last dial as making the "learning go faster," as making the parts smarter or more intelligent. Let's keep with our pattern: Let's start out with no learning; no adaptation. We have simple parts that follow fixed rules. We'll see in a few lectures when we talk about something called the game of life that it's possible for simple fixed rules to create complex behaviors; however, it's not likely. Note that we encountered this same insight when we talked about diversity and complexity. With fixed rules in our chemistry lab and in May's computer model, when we mix too much stuff together, we don't get complexity.

A little learning or adaptation goes a long way. It allows the parts to figure out how to interact with one another and how to create a coherent, complex whole. At least it can. What happens, though, if we crank the learning dial all the way to the right? The answer is: often we get equilibrium. We can see this as follows: Suppose we didn't get an equilibrium; suppose the system stayed in flux. Then it would be the case that some entity would want to change what it's doing. But if that's the case, the entity if it's super smart should have already changed what it was doing. Let's allow our entities to be people. If everyone's super intelligent and we're all optimizing relative to what everybody else is doing, then the system should be in equilibrium; because if it wasn't in equilibrium, someone should have done something different, and since we're super smart they already should have done it.

This isn't saying that everyone will best respond in the next period as in our game of rock, paper, scissors; this is saying that at every moment, everyone is optimizing. In game theory, this is known as a Nash Equilibrium. This is a core assumption of neoclassical economics: that people optimize. If people can optimize, we don't get complexity, we get equilibria. This why economists believe markets are in equilibrium, because they think that people are smart enough to solve the problem, and if everybody can solve their problems, the system's going to equilibrate. We'll see later on when we talk about negative feedbacks in a later lecture why this isn't necessarily a bad assumption for markets; however, it's not a good assumption in all cases. For the moment, I just want to focus on this idea of why ratcheting up sophistication leads to equilibrium.

Indulge me in a quick story; this is about the late Marion Tinsley and Jonathon Schaeffer. Schaeffer's a computer scientist. Tinsley was a mathematician and he's considered the greatest checker player in history. He was world champion 20 times, and he's reputed in his entire life to have lost less than a dozen games; this is 45 years of playing checkers, this guy lost less than a dozen times. Tinsley spent something on the order of 10,000 hours in graduate school studying checkers.

A good game of checkers is complex; each person's adapting to the moves of the other players, and, as a rule, the smarter the players, the more complex the game. By the way, the reason checkers supports so much complexity is that it has lots of possible plays: there are 500 billion billion possible plays of the game of checkers to be exact. So enter Jonathon Schaeffer. Schaeffer, with a group of coauthors, solved checkers. That's right: In 2007, they wrote a paper published in *Science* with the title "Checkers is Solved." What this means is that checkers is no longer complex; we can just look at their solution and follow the rules. That solution will be an equilibrium. That equilibrium, by the way, is a draw; it's just like Tic Tac Toe. What this means is if you're really smart, checkers is Tic Tac Toe, it's not at all complex. Checkers is only complex if you're in that "interesting in between" region; if you're reasonably smart. The same could someday be true with Chess; someone might write a paper saying "Chess is solved." What is now a complex game—Chess—is really tied only to our own limitations; once we become smart enough, it could become the equivalent of Tic Tac Toe.

So what have we learned? We've learned that when we say a system is complex, what we mean is that it produces interesting non-periodic patterns and emergent structures and functionalities; it's not a simple orbit, it's not chaos, there's structure. In this lecture, we've started to explore how varying the characteristics—the attributes—of a system make it more or less complex. We've found that complexity tends to lie in the middle region where we have moderate levels of interdependency; moderate levels of connectedness; we have some diversity but not too much; and things learn and adapt but they're not super smart. If we adjust these dials too far to the left, nothing interesting happens; and if we adjust them too far to the right, we basically tend to get equilibrium, or for completely different reasons we get a mangle. For example, if we have too much connectedness, every entity responds to an average that isn't complex. Or, as we just saw, if we turn learning way up, the entities tend to produce an equilibrium.

Implicit in what we've learned is an explanation of why we see so much complexity in the social world; and we'll talk about this in the next lecture as well. People are connected through networks both spatially and geographically; but we don't interact with everyone. We also have modest degrees of interdependency; we care what other people do, but it's not the case that every time someone we know changes a behavior that we feel compelled to change as well. Nor is it the case that we're immune to changes in the behavior of friends and coworkers. We're in that "interesting in between." We're also diverse. We're very diverse, but we're not that diverse. We eat similar foods, we pursue broadly similar goals, and we carry around roughly the same bag of tricks as everybody else. Finally, we're smart; we adapt. It's that adaptation that keeps things sort of churning. But we're not that smart. It took a team of computer scientists with an enormous computer to solve checkers; checkers! We haven't even solved chess yet.

Contrary to what economists assume, we're hardly up to the task of optimizing in all circumstances. What we do is we apply rules that we think make sense, and we adapt those rules when they fail to perform.

In sum: We're connected; we're interdependent; we're diverse; and we're smart; but we're not at the extremes in any one of these, so the world we construct is necessarily complex. In our next lecture, we're going to dig a little bit deeper into the other roles that diversity plays in complex systems. For not only does it produce complexity, we're going to see that it produces robustness and novelty.

Why Different Is More
Lecture 4

With just two distinct bits—a zero and a one—and enough time, we can produce all the differences that have ever been and that ever can be.

W e now focus on the role of difference—either variation in type or diversity of type—in complex systems. It is important to realize that a population that is referred to as a unified whole can differ genetically and phenotypically. This variation within a population allows adaptation. Before we can talk about diversity, we have to know what it is. Over the past half-century, statisticians, ecologists, computer scientists, and economists have proposed a variety of diversity measures. We can distinguish between four types of diversity measures: variation measures, entropy measures, distance measures, and attribute-based measures. Measures can either be constructed from the ground up by experimenting with mathematical formulae, or they can be derived analytically from a list of desiderata. Diversity measures compress information. They transform sets of diverse entities into single numbers. In the process, meaningful distinctions disappear. Because diversity measures can be applied to a variety of entities, and because each of these sets has distinct properties, we should not expect a one-size-fits-all measure.

Library of Congress, Prints and Photographs Division, LC-DIG-npcc-14348.

The great inventor Thomas Edison experimented with combinations.

Variation measures capture differences along a single numerical attribute. We capture that variation by means of a distribution, which plots the range of values and their likelihood. The most common measures of this type are statistical variance and its square root, standard deviation. To take into account different types, we need an entropy measure. Entropy measures capture the evenness of a distribution across types. Entropy measures depend on the number of types. The more types, the more entropy. The flaw with entropy measures is that they do not take type-level differences into account.

Two types of measure have been constructed that do take type-level differences into account: distance measures and attribute measures. Distance measures assume a preexisting distance function for pairs of types—for instance, genetic distance in the case of species. Attribute measures identify the attributes of each type in the set and then count up the total number of unique attributes. There are four causes of diversity in complex systems. Perhaps the biggest cause of diversity is diversity itself. The more diversity you start with, the more you can produce. The second cause of diversity is weak selective pressure. If there is no selective pressure, nothing stops diversity from spreading. Another cause of diversity is different landscapes. In this case, the diversity arises because the problems that need to be solved are different, so the peaks differ. The final way in which selection can produce type diversity is through dancing landscapes, where movements on one landscape shift the heights on other landscapes.

Diversity measures compress information. ... In the process, meaningful distinctions disappear.

Can we say anything about how systems in which diversity evolves, like ecosystems, differ from systems in which diversity is created by purposeful actors? Let's look at a few key differences. The first difference relates to the size of the leaps. Evolution is a plodder, but creative systems can take big leaps. The second difference relates to interim viability. Evolution is constrained in that each step along the path to an improvement must be viable. Creative processes do not have this constraint. The third difference relates to representation. Evolution is stuck with genetic representations. It cannot switch to some new encoding. This is not true for creative systems. ∎

Suggested Reading

Page, *The Difference.*

Weiner, *The Beak of the Finch.*

1. Suppose that you run a research-based organization. How might you know if you have too little or too much diversity among your researchers?

2. Can you think of an example of where less variation would be better? Does this example contradict what we learned in this lecture?

Why Different Is More

Lecture 4—Transcript

In this lecture, we focus on the role of difference—variation in type or diversity of types—within complex systems. I want to start, though, by telling you a little bit about Indigo Buntings. If you were to pick up a bird guide, it will say something like the following:

> The Indigo Bunting, identifiable by its males' dark blue feathers, lives on the edges of farms and roadways and in the brushy open spaces beneath power lines. Though small birds, male Indigo Buntings produce loud, piercing warbles. Neighboring Indigo Buntings vocalize in near identical patterns, but Indigo Buntings separated by even a half mile sing different tunes.

If you saw a bird that matched that description you might say to yourself, "Look, that's an Indigo Bunting," but technically that wouldn't quite be correct. You should say, "This is a member of the population of birds referred to as Indigo Buntings." Why the hairsplitting? Why make this distinction? The hairsplitting is necessary because the population of birds called Indigo Buntings differs; there's variation both genetically and phenotypically. The birth or death of any one Indigo Bunting changes the definition of what an Indigo Bunting is. The fact that Indigo Buntings are a population and not a specific thing has profound implications. This variation is going to enable adaptation. For example, suppose there was a decade-long drought and the only food source that Indigo Buntings had required shorter beaks; they had to get their beaks in these little crevices—or these wide crevices—and break little nuts or something. What would happen, then, is for the Indigo Buntings to survive those with shorter beaks would do better; and within a few generations, the whole population of Indigo Buntings would have shorter beaks. As a result, what is an Indigo Bunting will have changed.

We can make similar statements about cultures. There's no such thing as a French person, per se. Instead there's a collection of people we call French, and these people differ genetically, and they have different talents, beliefs, preferences, and skills.

In this lecture, we're going to talk about variation and about diversity—they're not going to be the same thing—and we're going to see why both matter in complex systems. The lecture's going to have three parts. In part one, we're going to ask, what is diversity, and how do we measure it? We're just going to sort of figure out what it means exactly. In part two, we're going to learn why complex systems produce diversity; why they do it and how they do it. In this part, we're going to take a little detour that's going to be some fun; we're going to talk about differences between creative systems—human systems in which we create things—and evolutionary systems, and how those two types of systems produce diversity. In part three—the third part—we'll talk about the roles diversity plays in complex systems. Let me give away the plot just a tiny bit: The ability of complex systems to produce and maintain diversity enables them to be innovative and robust.

Part one: the naming of the parts. Before we can talk about diversity we have to know what it is, and this isn't as easy as it sounds. Over the past half century, statisticians, ecologists, computer scientists, and economists have proposed a variety of diversity measures. We're going to distinguish between four types: measures of variation, entropy measures, distance measures, and attribute measures. Don't worry if you don't know what these words mean; we'll get to them in a second.

Before discussing them, I want to make three observations about measures in general and diversity measures in particular. First, measures either can be constructed from the ground up by experimenting with mathematical formulae, or they can be derived analytically by putting down a list of desiderata; these are sort of axioms (we want our measure to satisfy this). The latter approach appears more scientific, but in practice the two approaches are pretty much the same. Most intuitive measures are going to prove to satisfy some axioms; if they didn't, the measures wouldn't be useful. Therefore, any measure that's presented with some sort of great authority as the unique measure or class of measures that satisfies these four axioms might well have been derived from intuition and then bolstered by the logic.

Second, diversity measures compress information. They transform sets of diverse entities into single numbers; into one number. In the process, what's happening is we're really condensing information; information is getting lost. For every one of the measures I'm going to describe, if you sit back and think about it a little bit, with some effort you're going to realize that we can take sets of things that differ fairly substantially, yet they're going to have the same diversity measure. For example, we might take one of these diversity measures off the shelf and find that the diversity of fonts in Microsoft Word equals the diversity of deciduous trees in Ohio. That's just nuts, right? But it could be true. It also, though, has some serious implications. Suppose we wanted to see if more diverse ecosystems were more robust; so we wanted to do an empirical test. We could gather data and go test the claim. But if we do such a test, it's going to suffer from this problem of information condensing. Such a test would treat species of frogs and species of trees the same way; this isn't as goofy as treating fonts and trees the same way, but it's still a large leap (no pun intended).

Finally, because diversity measures can be applied to a variety of entities—cultures, languages, makes of automobiles, even toothbrushes—and because each of these sets of types has distinct properties, we should not expect a one-size-fits-all measure; so we're going to have lots of measures. Let me flesh this out a bit. Bird species have a genetic lineage, and they can be mapped into a graphical topology that has branches as we move back the evolutionary tree. Measures of bird diversity can exploit this branching structure, but this same measure won't work for breakfast cereals because breakfast cereals have no genetic history, so they can't be arranged in a branching network. The abundance of measures isn't going to be a bad thing; it's going to be a good thing. It's going to allow us to pick the measure that fits the context. It's also going to allow us multiple lenses to look at the same entities.

Let's look at the measures. The first type of measures: these are variation measures; they capture differences among a single numerical attribute. An example if we go back to the Indigo Buntings would be something like beak length: Different Indigo Buntings have different lengths in their beaks. We can capture that variation with a distribution. What a distribution does is plots the range of values and their likelihood. A bell curve, a normal distribution that we'll talk about later, is an example of a distribution.

Let's take an example: Consider a gas. Within a gas, each particle has a velocity and these velocities differ. Gas velocities are distributed according to something known as a Boltzman distribution. A Boltzman distribution looks sort of like a bell curve but it's sort of smooshed up on the left-hand side and pulled out on the right-hand side; so there are not as many slow particles as you would like, and there are a few more fast ones. This variation in particle velocity matters. The gas molecules that escape our atmosphere—the same ones that put holes in the ozone layer—are the ones that move fastest.

How do we measure variation? The most common measures are what we call statistical variance, and its associated square root standard deviation. How do we compute this? Variance is just the average squared distance to the mean. If I have two Indigo Buntings, one weighs 4 ounces and the other weighs 8 ounces, the average weight will be 6 ounces. Each of these two Indigo Buntings differs from the mean by 2 ounces ($6 - 4 = 2$, and $8 - 6 = 2$); I square those two things ($2^2 = 4$); so in this case the average square distance [$(4 + 4)/2$] is going to be 4.

The question to ask here: Why are we squaring things? We aren't I just taking two plus two? The reason we square is we want to make all the differences positive; so if we have plus two and we have minus two, if we added those up we'd get zero, but by squaring them first I make sure that variation is positive.

Variance is going to be great if we can assign numerical values, but often times we can't. If I have 40 tables sitting in front of me and they're all in different colors, and I want some measure of their color diversity, there's no way of assigning numbers to those colors that makes any sense. To take into account the fact that I could have different colors or different types, what I need is an entropy measure. Entropy measures capture the evenness of a distribution across types. If red and blue tables are equally likely, then entropy is going to be higher than if red tables are twice as likely as blue tables. What entropy measures do is they depend on the evenness of the type and the number of types; so the more colors of tables I have, the more entropy I have.

The best way to really get a handle on an entropy measure is to actually do some examples. We're going to consider one specific example, and this is called in biology the Simpson's Index, political scientists call it the effective number of parties, and economists call it the Herfindahl Index of market concentration. Despite these many names, it's the same measure.

Let's go back to the multicolored tables. Take the proportion of each color of table and square it; that's what we want to do. For example, if a third of the tables are red, we square that and we get a nine. What we do is we take each of these proportions, we square them, and then we add them all up; that's step one. Then step two, we take the inverse; we just take one over that number. Let me do an example: Suppose there's $\frac{1}{3}$ red, $\frac{1}{3}$ white, and $\frac{1}{3}$ blue tables. So what I'd do is take $\frac{1}{3}$ and I'd square it and I'd get $\frac{1}{9}$, and I'd do that three times; so I'd get $\frac{1}{9} + \frac{1}{9} + \frac{1}{9}$, that's $\frac{3}{9}$ or $\frac{1}{3}$. Then I'd just take the inverse of that and that gives me 3. Notice I had three tables of equal likelihood, and this diversity measure gave me three back. It turns out if I had 5 types of equal likelihood and I did the diversity measure, I'd get 5 back, and if I had 10 types I'd get 10 back; in fact, if you had any number of types in equal amounts, this diversity measure—Simpson's Index; Herfindahl Index—the diversity's going to equal the number of types.

It's also going to be true—we're not going to bother with the math—that if you have three types but they're not evenly distributed or spread (so $\frac{1}{2}$ are red, and $\frac{1}{4}$ each are blue and yellow), the diversity's going to be less than 3. In fact in that case, the diversity would be 2.6. This seems like a great way to measure diversity, right? If we have N types equally spread, we get a diversity of N. As they become less evenly spread, diversity falls. These are the sort of axioms, or desiderata, that we'd want a diversity measure to satisfy. This seems great, right? As good as this measure is, it has a flaw; it has a fundamental flaw, and that flaw is this: It doesn't take into account the differences between the types. For example, if I have a basket with 10 apples, 10 pears, and 10 oranges, that's going to have more entropy or more diversity than a basket with 11 apples, 10 iguanas, and 9 orchids. The reason why is the first basket has a more even distribution; however, all that fruit— the apples and oranges—are more similar than the apples and the iguanas. Entropy measures fail to take into account that iguanas and orchids differ from apples much more so than pears and bananas.

As difficult as comparing apples and oranges may be, comparing oranges and iguanas seems a little bit harder. For that reason, we have to introduce two other types of diversity measures that take into account differences between the entities. I'm just going to do these quickly and then move on to other stuff, but these are important when we talk about the creation and the functions of difference.

Distance measures: What these do is they measure a preexisting distance function between pairs of types. Remember when I talked about how genes let us tell how far apart two species are with a branching back in time? That's what a distance function measures. We can use this to show that an apple is closer to a pear than it is to an iguana. If we have this distance function—and we may not—one way to measure diversity would just be to add up all the distances between the members of the set and take an average. This approach won't be perfect, but keep in mind the members of a set have distance between them; the more distance that you see, the more diversity there's going to be. This idea is going to be important when we talk about the creation of diversity.

Attribute measures are the second way to identify type level differences. What these do is they identify the attributes of each type in the set and then count up the total number of unique attributes. Example: Suppose I have a frog, a table, and an elephant. I could list all the attributes that a table has: it's wooden, it has a flat top, it has four legs, and so on. I could do the same for a frog: living creature, four legs, swims, etc. Notice that all three—the frog, the table, the elephant—have four legs. You might say, "Wait, why would somebody care about them having four legs? Why would I care about attribute diversity? The simple reason is this: coverage. If I'm putting together a team of people to take on a business trip to Europe, I might want to make sure that I have at least one person who can speak the language of every country we're going to visit. I might not care at all about the entropy of the distribution of their language skills; entropy's not going to get us a hotel in Poland. I need someone who speaks Polish.

Now that we have some understanding of what these diversity measures are, we're ready to get some idea of how diversity gets produced in complex systems. I'm going to describe four causes, and these causes are going to hold both for creative systems and evolutionary systems. The first cause introduces an idea that we're going to discuss at length in a later lecture: positive feedbacks. The second has to do with how competitive or harsh the environment is. The last two are going to trace back to our ideas of rugged and dancing landscapes.

Cause one: diversity begets diversity; this is a positive feedback. This may be the biggest cause of diversity: diversity itself. The more the diversity that exists out there—whether it can be measured in variation, entropy of types, or attributes—the more diversity that we can create. Think of it simply: Suppose you have a box of Legos sitting there, and all you have are red two by two blocks, blue two by two blocks, and white two by two blocks. You can make some things—you can make a candy cane, a barber pole—but you're going to have a heck of a time making a tree or a car. Once I give you hundreds of different shapes, sizes, colors, and pieces, you can make just about anything. In fact if you go to Legoland, you can find a miniature version of New York that has the Empire State Building and the Statue of Liberty. There's no way you could make those things with a few regular pieces of red, white, and blue Lego.

The same is true with species. If you look at the tremendous variation in the types of dogs that we have—Beagles, Mastiffs, Pomeranians, and even hybrids like Labradoodles—we'll see that what once we get this diversity of dogs we can mix them to create even greater diversity. The logic here's totally straightforward: the more diversity you start with, the more diversity you can produce.

Second cause: weak selective pressures. Biologists—not Darwin, by the way, but modern biologists—talk a lot about the survival of the fittest. When selection is not very severe, rather than survival of the fittest, we really have death of the unfit. If there is survival of all but the least fit, then diversity is really going to spread. To experience this firsthand, next time you're at the mall or the airport, look about you at the amazing diversity of attire that you're going to see on people. You're going to see something like a woman wearing brown shoes, purple pants, and a tan coat maybe next to some guy in blue chinos and a crisp striped shirt. The amount of diversity, no matter how you measure it—entropy in number of colors or styles, or some sort of attribute measure—is mindboggling. But if you walk into a law firm or if you walk into an investment bank, you're going to see very little variation in the way people dress. Why?

In the first case, there's not a lot of selective pressure; there's no selection based on how we dress during leisure time. No one asks you to leave the mall or the airport if your socks don't match or if you're wearing a Clapton t-shirt that doesn't match your chinos. That's not true in the world of business. In business, attire matters a lot. That's why books like *Dress for Success* consistently sell: If you don't dress a certain way, you're not going to get a promotion. If there's very little selective pressure, nothing stops diversity from spreading.

Cause three: This is the first of our sort of landscape causes. If we have different landscapes, we're going to have a lot of diversity; so let's go back to the mall and all that diversity we see in how people dress. To the extent there's any selective pressure at all in how we dress, it's different. Most of us, at some level, want to look cool or at a minimum sort of have a look; but the selective pressure we face to be cool differs. Someone who wants to look Goth—that's a look that sort of requires wearing a lot of black—has different selective pressure than someone who's trying to maybe look like they're on a route to success or trying to get a job.

One of the most famous examples of different landscapes comes from Darwin himself and has to do with finches. Darwin showed that finches that confront different environments evolve beaks with different shapes and sizes. If a finch has to poke holes to get food, it develops a long narrow beak. If it has to crush seeds, it evolves a short, powerful beak. This diversity arises because the problems that evolution is giving them—the problems they have to solve—differ, so they're climbing different landscapes, so their solutions differ.

In some instances the problem being solved can be the same, but evolution or the marketplace still produces diversity. This happens if the landscape has multiple peaks—if we have a rugged landscape—and different experiences on that landscape find different peaks. Multiple peaks on the same landscape aren't quite the same as climbing different landscapes, but the net result is the same: diversity.

Last cause (it shouldn't be a surprise): dancing landscapes; complexity. This is the final way in which we get diversity, and again it's caused by selection. Recall if we have a dancing landscape—or maybe another word used for this is a coupled landscapes—we have interdependencies in payoffs. The fitness of a worm depends on the characteristics and behaviors of birds and other predators. The same is true in an economy. The profitability of an airline depends not only on its routes, pricing, and fleet, but also on the characteristics and actions of its competitors. Metaphorically, remember, this means that landscapes perform a coupled dance; we talked about this in an earlier lecture. Movements on one landscape shift the heights of the other landscapes. When the gazelle become faster, the fitness of extra bulk on a lion falls. Or, when technology improves the dependability of package delivery, the benefits of creating an online purchase option increase. If your competitor changes its strategy, this may change what you do; the coupling of these dancing landscapes produces different outcomes.

We've discussed how complex systems produce diversity: diverse parts, weak selection, different landscapes, and dancing landscapes. All four of these can be applied to ecological as well as social systems. That's a really cool thing about complex systems: they apply broadly; they fan out. But how broadly? The market isn't an ecosystem, and an ecosystem isn't a market. What we want to do is we want to see how these two sorts of systems differ. Let's just look at a few key differences. The first difference refers to the size of the leaps. For the most part, evolution is a plodder; we can't cross a leopard with an ostrich (at least not yet). But that's not true in creative systems; creative systems can take huge leaps. We can take a light bulb and a helmet and cross them and get a mining helmet.

Second difference: interim viability. Evolution is constrained in that each step along the path has to be viable. The evolution of the human eye occurred through a sequence of steps, each of which was viable. The same is true of the four chambers of our heart. These chambers could only have evolved if each step in the process creating them contributed to a viable entity. One theory, for example, proposes that the heart evolved from a linear tube that wrapped around on itself to create parallel chambers. Notice how this contrasts with creative processes where interim viability just isn't a concern. If you talk to someone who designs products or any sort of engineer, you're likely to hear her say, "We're close; it's still not working, but we're close and we think we see a way to solve this," or something like that. Evolution has no such luxury; it can't be close.

Third difference (we've talked about this before): representation. Evolution is stuck with genetic representations; it can't switch to a new encoding. That's not true for creative systems. We talked earlier about how there was no rugged landscape in creative systems. Often, a creative act consists of applying a new representatIon to an existing problem.

Fourth: The fourth difference is retrievability. In creative systems, we can go back in time and resurrect old ideas and products. The Volkswagen Beetle was out of production for more than a decade, and then it's resurrected; albeit with some modifications, but it was resurrected. *Jurassic Park* notwithstanding, that's not possible in evolutionary systems. True, DNA contains history of past traits, and it's true it's possible evolution can go back to a previous design, but an evolutionary system can't just open up a catalog and order up a Woolly Mammoth. The Woolly Mammoth is gone.

These four reasons so far suggest that creative systems should be way more diverse than ecological systems; but we have to hold on here, let's not move too fast. We've left out one thing. Remember the story of the tortoise and the hare; remember the moral? Slow and steady wins the race. Evolution is relentless; it is slow and steady. It never, ever gives up. It keeps testing and trying. It tries anything—at least anything that's close—in the genetic record. That need not be true in creative systems; we're often blinded by what are called dominant logics. Problem solvers may be blind to some idea; even though people can combine almost anything with anything, we sometimes don't.

Example: 1843, Charles Goodyear received a patent for vulcanization; this is the process through which you remove the sulfur from rubber. When you vulcanize it, rubber became more elastic, waterproof, and weather proof; it becomes like modern rubber, it becomes useful. The world had to wait 12 years before someone named Stephen Perry came up with a rubber band; 12 years! And we had to wait 56 years—until January 24, 1899—before a man named Humphrey O'Sullivan thought of putting rubber soles on shoes. The parts were there the entire time—shoes and rubber—but it took 56 years. What a creative person combines and/or mutates depends upon the creator's context; their milieu. Ideas that in another time or place may seem obvious may not be considered because they don't belong to the set of ideas that are inside our heads. Evolution is going to plod its way through these things and try anything; humans in creative systems try what they think of, and that can lead to blind spots.

We now have some idea of what diversity is, how it gets created, and how ecological and evolutionary systems deal from creative systems; now what we want to do is we want to see why diversity matters. We want to focus on two main contributions of diversity in complex systems. First, I want to talk about how diversity produces innovation. Let's think back to how diversity begets diversity; how with more parts—more pieces of Lego—we can produce more stuff. Successful innovation requires sorting among all that stuff and choosing the best. Let's see how this works.

Let's turn to maybe one of the greatest inventors of all time, Thomas Edison. In his lifetime, he was awarded 1,093 patents. He invented among other things the phonograph, the light bulb, and the motion picture camera. At Greenfield Village in Dearborn, Michigan (near where I live), you can go visit Thomas Edison's old laboratory. Henry Ford literally had it shipped board by board brick by brick from Menlo Park to Michigan. On the shelves of Edison's laboratory (this is totally cool) you can see jar upon jar of chemicals, herbs, metals, and so on. These were his building blocks; these were his Legos.

Imagine there were only 100 such jars; there were many more, but let's just suppose 100. The number of pairs of jars that you could try is (100 × 99)/2. Why is that? There are 100 jars to choose from first, and 99 to choose second; hence, 100 × 99. However, if you were to choose shoes and then rubber, that's the same thing as choosing rubber and then the shoes, so we have to divide by 2; so hence, (100 × 99)/2. That's roughly 5,000. We can do similar math, and we can find out that there's 161,000 ways to pick three items, and there are 3.9 million ways to choose four jars, and 125 million ways to choose five jars. That's a huge number. So once we start creating stuff, we get this combinatory explosion of new things that can be created. Edison worked by experimenting with these combinations. He once said, "We now know a thousand ways not to make a light bulb." But he had a 125 million ways to make a light bulb, and so he found one that worked.

Thinking about these 100 jars and these 125 million ways to select five of them, it's important for us to return to our measures of variation and diversity. We're not talking about variation here; what we're describing is diversity of types, which is measured by entropy. We're also sort of implicitly assuming some distance-based diversity or some of attribute-based measure depending on the items in the jars. If they weren't different in their attributes, putting them together wouldn't create anything interesting.

Diversity drives innovation in one way by creating lots of opportunities; it also drives it in a second way. If we allow the parts to be less tangible—to be things like ideas, representations, and thoughts—then it's possible to show how diversity improves problem solving. Here's why: Suppose we have someone who's stuck on a local peak; let's go back to our rugged landscape. Someone else represents the problem differently, or someone knows another direction to go; instead of just north or east, they know to go northeast. If that's true, what's a peak for the first person may not be a peak for the second person; as a result, the second person can find an improvement. History is full of such examples—we talked about this in the past—Einstein's theory of relativity and Mendeleev's formulation of the periodic table; these are cases where someone represented a part of the world differently and made a breakthrough. Or we have other examples: Newton coming up with a calculus; this is a case of someone coming up with a new way of searching the landscape and getting us to new peaks.

In addition to being a source of innovation in complex systems, diversity can also contribute to robustness; this will be our second big feature of diversity in complex systems. In a later lecture, we're going to discuss something called positive and negative feedbacks; we already talked about positive feedbacks a little bit here where diversity beget diversity, and how when feedbacks are negative we're going to see how diversity can contribute to stability. Here we're going to talk about models without any feedbacks, and we're going to see how diversity creates robustness.

First, I want to see how variation produces robustness. What do I mean by robustness? I mean the ability to maintain some sort of functionality despite a disturbance. Suppose you have a screw loose—not in your head, but at home—and you only have a single screwdriver, it may be that the shaft is too short, or the blade is too wide, or it's too narrow to fit in the screw head or something, it might not work. But if you have a lot variation in your screwdrivers—say a set of 12—you're bound to have one that fits. Variation in your collection of screwdrivers enables you to respond regardless of which screw is loose.

We've already talked about this, but in evolutionary systems genetic variation plays this same role. Recall the opening to this lecture, that there is no Indigo Bunting per se, but there is a population of birds that we call Indigo Buntings and that population has variation. It's that variation that enables the population to be robust to environmental changes; the differences in beak length and beak strength.

Variation clearly matters; what about diversity itself? Does it enhance robustness? Yes, absolutely. Let's go back this analogy of toolboxes. If we have more types of tools—hammers, saws, screwdrivers, even axes—then we're more likely to be able to solve or cope with any problem. The same is true in an ecosystem: The more species, the more likely that some—or at least one—is going to survive some sort of change in the environment, say a climactic change. Example: If your garden consists of only roses and the year is very, very wet, then it's going to be the case that rose bushes may all rot. But if your garden has many varieties of plants, it's likely that at least some of them are going to survive. The idea that the more types there are, the more responses you get, and the more robust the system is underpins the concept of requisite variety. Requisite variety says that robustness requires for every disturbance there must exist a response. If it rains, you need an umbrella; if it snows, you need a shovel; if you have a screw loose, you need a screwdriver. This makes sense.

But there are bigger fish to fry here, literally. Recall our discussion about Robert May's modeling showing how diversity reduces robustness, and the contradictory empirical evidence. The fact that the relationship between diversity and robustness is an open question hints at the limitations of these sorts of one-dimensional characterizations. Complexity has many dimensions and many definitions, as does diversity, as does robustness. To say diversity improves robustness in complex systems is to speak very crudely, and what we need is more careful thinking and more innovative modeling. In coming lectures, we're going to see some research that moves in that direction. That's going to be particularly true when we talk about feedbacks; we're going to see how diversity plays both roles.

For now, though, let's just exhale, rest a bit, and revel—intellectually at least—in the challenging questions that are before us, and we'll try and contemplate as we move forward this interesting and challenging relationship between diversity and complexity. In the next lecture, we're going to look at another aspect of complexity: the question of balancing the search for better solutions against the likelihood of finding them.

Explore Exploit—The Fundamental Trade-Off
Lecture 5

A complex system consists of these little entities that have interdependent payoffs and rules that create emergent phenomena, robustness, and possibly large events.

A fundamental trade-off in a complex system is exploration versus exploitation. By exploration, we mean searching for better solutions. By exploitation, we mean taking advantage of what you know—reaping the benefits of past searches. Ideally, exploration should be balanced against exploitation. Why must actors in a complex system maintain this balance, and how does doing so help maintain complexity? We begin by describing the explore/exploit trade-off in the context of a decision problem called the two-armed bandit problem. We see how the explore/exploit trade-off manifests itself in rugged and dancing landscapes. We discuss an algorithm for balancing exploration with exploitation called simulated annealing. We then turn to evolutionary systems and see how the basic mechanisms of evolution can be seen through the prism of the explore/exploit trade-off.

The two-armed bandit problem. Imagine a slot machine with two levers, one on the right and one on the left. The levers offer the same payout but at a different rate. The problem is how to balance exploration (testing the rates of both levers) and exploitation (acting on a determination based on previous testing) to maximize return.

We return to the concept of a rugged landscape and describe how exploration and exploitation play out there. Recall the hiker whose ambition is to find a point of high elevation. This hiker has limited time and wants to spend as much of it as possible on high ground. Using the rule that he only proceeds with each step if it takes him higher, the hiker would very quickly reach the summit of a Mount Fuji landscape. However, this rule would perform poorly in a rugged landscape. Computer scientists refer to this sort of search rule as a greedy algorithm.

A much better alternative would be the search approach known as simulated annealing. We study simulated annealing for three reasons. It works. It provides a segue into the concept of self-organization. It shows the fan-out nature of complex systems. First of all, what is annealing? Annealing is used to harden glass and metals and to make crystals. Spin glass is a stylized model of glass, metal, or crystal. It is an enormous checkerboard of particles, each with a spin (either pointed up or down). The goal is to get all of these particles to point in the same direction.

Glassmakers and metallurgists use annealing to accomplish this. Annealing takes advantage of the fact that particles want to line up with their neighbors, much like magnets. When the metal is too cold, the particles are frozen in what physicists call a frustrated or disorganized state. The trick seems to be to heat the metal until the particles are free to move; but if we keep the heat on high, the particles will never settle down. If we heat the metal just enough so that the particles can move, they will align with their neighbors, but not all local neighborhoods will agree. The result will be a camouflage pattern. Once the camouflage pattern forms, we cool the temperature a bit to achieve what is called annealing. The boundaries of the neighborhoods shift back and forth until by chance the regions absorb one another and become uniform. Once this is achieved, the temperature is cooled further so that the system freezes in an organized state.

Natural selection is a form of exploitation, resulting in genetic fitness.

Let's go back to our hiker and apply a simulated annealing algorithm to his searches. Think of temperature as a proxy for the probability of making a mistake. If the temperature is high, lots of mistakes are made; if very low, none. In this model, a higher temperature means more exploring, and a lower temperature means more exploiting. When we put our hiker in a rugged landscape, the temperature starts high—he just explores a lot. As we cool the temperature, the hiker becomes more likely to go up than down. When the temperature becomes very cool, he inevitably ends up on a local peak. While this may not provide the optimum solution, it does provide a good one. Optimal cooling depends on the ruggedness of the landscape.

Evolution balances exploration and exploitation. Mutation and recombination of genes are forms of exploration, resulting in genetic diversity. Natural selection is a form of exploitation, resulting in genetic fitness. Although this is a gross simplification, it is a useful one.

Let's now turn to dancing landscapes—in particular, the complex system of leafcutter ants. Leafcutter ant colonies are massive productive systems, with ants of various sizes each suited to a different task. How is this a dancing landscape? The fungi the ants produce attract bacteria, which threaten to overwhelm the ants. The ants have an antibiotic system consisting of different bacteria that grow on their backs, which attack the invading bacteria. This shows a constant balance between exploration and exploitation. The leafcutter ants had to form a relationship with the bacteria in order to maintain elevation on the dancing landscape.

Therefore, rugged and dancing landscapes require different explore/exploit balances. We saw how on a rugged landscape the balance between exploration and exploitation should end with almost complete exploitation. Hence we devise algorithms like simulated annealing, which start out exploring but end up exploiting. On dancing landscapes, agents can never stop exploring. This explains why we see complexity: Equilibrium allows for exploration, which stimulates dancing landscapes. Randomness is avoided because as exploration becomes prevalent, the value of exploitation increases. Thus individual agents balance the necessity to explore and exploit, producing complexity as a result. Complexity can thus be thought of as an emergent property. ∎

Suggested Reading

Kauffman, *At Home in the Universe.*

Questions to Consider

1. What do you make of the following claim? "Organizations that attempt to turn dancing landscapes into rugged landscapes ultimately fail."

2. How might an organization apply the concept of simulated annealing into its standard operating procedure for developing new policies? Recall that the temperature can be thought of as the probability of making a mistake.

Explore Exploit—The Fundamental Trade-Off
Lecture 5—Transcript

Remember that a complex system consists of these little entities—these agents, these parts—that have interdependent payoffs and rules that create emergent phenomena, robustness, and possibly large events. In this lecture, we're going to talk about a fundamental trade-off for the agents within a complex system, and that trade-off involves whether to explore or whether to exploit. First, some basic definitions: By "exploration," I mean searching for better solutions (climbing on that landscape). The more you search, the more likely you'll locate a really good choice of action. By "exploitation," I mean taking advantage of what you know; reaping the benefits of past searches. This trade-off between continuing to search (to explore), or to rely on what you've already learned (to exploit) is common for actors within a complex system.

Suppose, for example, you want to find a deli in New York. New York has thousands of delis. In addition, new delis are popping up all the time, and existing delis adapt; they change their menus and suppliers. If you desired, you could explore forever; but if you did this—if you kept exploring—on average, the quality of that corned beef on rye that you're eating would be just average. At some point, what you have to do is exploit the information that you've gathered. But if you stop exploring—if you go the other way, if you just exploit what you know—you could miss out on something much better; or even worse, your loyalty and that of others could cause the quality of your favorite deli to decline, and what was once a great place to eat may now be just sort of mediocre. Ideally, you shouldn't either stand pat or continue to explore; what you need to do is balance exploration against exploitation.

In this lecture, what we're going to see is how and why actors in complex systems maintain this balance. In addition, we're going to see how doing so actually produces—maintains—complexity; but we're only going to get to that at the end.

Recall from a previous lecture that a system with very few connections and interdependencies won't produce complexity. Instead, it's just going to be a bunch of isolated independent events. But if we had too many connections and interdependencies, the result would be some sort of incomprehensible mangle, or grey goo. The domain of complexity exists in this "interesting in between" region, where we have some connections and some interdependencies, but not too much. That same logic held for diversity and adaptation: Complexity happens in the region in between. But this begs the question: If complexity happens when those dials—remember our four dials—are adjusted just right in the center, how do they get there? In this lecture, we're going to give a possible answer as to why.

We're going to begin, though, by describing this explore/exploit trade-off, because that's going to be the basis for this fundamental insight; and we're going to do it in the context of a rugged landscape problem: a decision problem called the two-armed bandit. This problem requires choosing between two arms on a slot machine. After we do that, we're going to return to two concepts from the second lecture—rugged landscapes and dancing landscapes—and we're going to see how the explore/exploit trade-off manifests itself in those two contexts. Along the way, we're going to introduce a really cool algorithm—that's just a search procedure—that balances exploration with exploitation, and it's called simulated annealing. This algorithm abstracts from a process used in physics and engineering— the annealing of glass and metals—and constructs an algorithm (a heuristic) that can be used to solve problems. We're then going to turn to evolutionary systems and see how the basic mechanisms of evolution—mutation, recombination, and selection—can be seen through this prism of exploration versus exploitation. Species in an ecology, just like firms in a market, confront this tension between explore and exploit. Finally, we're going to use this idea—this explore/exploit trade-off—to propose a possible explanation for why systems tend to be complex.

Let's go to the two-armed bandit problem. If you walk into a Las Vegas or Atlantic City casino, you're going to see hundreds if not thousands of people sitting there dutifully pulling the levers of slot machines. In some places, these levers have been replaced by buttons; but for imagery, I want you to think that people are pulling levers. Most slot machines have a single lever. These machines are designed so that on average you lose money—I hope that's doesn't shock anyone, but they are—hence they're called "one-armed bandit." You put in money, you pull the lever; you win some, you lose some. As the old saying goes, "You pays your money, you takes your chances."

Think of a slot machine not with one arm but with two arms. The key assumption here is going to be that these two arms pay out at different rates. To keep this experiment simple, I want to imagine that it costs a dollar to pull the arm and the machine pays out either $10 or it pays nothing. The only difference between these two arms is going to be the frequency with which they pay the $10. I'm going to give you $1,000 in seed money, and you have 1,000 pulls to make as much money as you can. How are you going to do this? How do you allocate your pulls across the levers? Or more appropriately, what should you do if you are optimizing?

What you should do is to take maybe the first 10 or 20 pulls and explore; so you might take 10 pulls of the right lever and 10 pulls of the left lever and compare the payoffs. If the right lever paid off three times and the left lever paid off twice, you might think, "Maybe I should pull the right lever more." But this would be exploiting the right lever: Do you exploit the information you've gathered so far, or do you continue to explore? 10 pulls isn't very many, and two wins is pretty close to three; so what you may do is decide to give 10 more tugs on each arm.

Suppose after these 10 more tugs, you have a total of six wins on the right arm and only four on the left arm. Now it might seem like a really good time to start exploiting; and you might be smart to allocate your next 10 pulls to the right arm. If again you found that 3 or more of those pulls pay off, you might be justified to stick with the right arm for your last 900-and-some pulls. But if you only get one or two payoffs in the next sequence of 10, you might decide, "I think I'm going to go back and explore a little bit more with that left arm."

Now you have the basic idea: The more you explore, the more likely you are to find the correct lever (that's a good thing); but the more you explore, the less time you spend taking advantage of your information (that's a bad thing).

The optimal solution to the two-armed bandit problem depends on the number of pulls you get and on the information you receive along the way. For example, if the right arm always pays off, you should keep pulling it forever; you might never even pull the left arm. Alternatively, if the right arm pays off only one in 20 pulls and the left arm pays off only one in every 25 pulls, you might spend most of your 1,000 pulls exploring and little time exploiting, because the difference is so small and the payoffs are so rare.

Now that we've got this basic idea down—this tension between exploration and exploitation—I want to take that basic idea and I want to go back to our concept of a rugged landscape and a Mount Fuji landscape and describe how that trade-off plays out there. Recall from the second lecture that we can think of the possible actions that an entity might take—this could be a frog, a person, or the FDIC—as a set of geographic coordinates. We're going to get a fitness of what that behavior is—the fitness of a species, or the happiness of an individual, or the success of a policy—in terms of elevation, and this gives us a landscape. Remember, the landscape can be either something like Mount Fuji, it can be very simple; or it can be rugged, like the Appalachians or the Rockies.

Let's go back to the idea we had of a hiker climbing on a landscape, searching for some point of high elevation. The ambition of this hiker is to get as high up as possible. If time weren't an issue, the hiker could check every point, explore forever, and then just choose the best point; that's easy. But the hiker, like us, has only so many grains of sand in his hourglass, so they're insufficient to perform this sort of exhaustive search. As the great philosopher Groucho Marx once said, "Time flies like an arrow. Fruit flies like a banana."

Given this limited time—given that time does fly—we can't search forever; we have to balance off exploration with exploitation. We have to learn what we can and then take advantage of what we've learned. The same is going to go for our hiker: He doesn't want to spend too much time searching around; he wants to get to high ground as quickly as possible.

So here we have our hiker, and let's suppose he takes his first step; he starts to explore. He's going to stay at this new point if this location is higher than where he was before. If the step leads him upwards, he's going to stay there; but if it takes him lower, then he'll return to where he was, because he's worse off. Starting from any point, the hiker has four directions to test: to the north, south, east, and west. If any one of those four directions lead him uphill, he'll take it. If none do, though, he's going to stop. Recall from the rugged landscape lecture that if he's at some point where north, south, east, and west are all worse, this is what we call a local peak; a local optimum. If our hiker uses this rule of only going up, what he's going to do is climb until he finds a local optimum, and once he finds one, he's going to stop.

Is this a good thing or a bad thing? Let's put him on Mount Fuji. If we put him on Mount Fuji and set him loose, he's going to clamber right up the side of the mountain to the peak. Excellent; he's at the global optimum, he's at the global peak, everything's fine. Let's put him in the Appalachians or the Rockies on a rugged landscape. In this case, this search rule is going to perform really poorly: He's going to climb the first hill he gets to, he's going to stand atop that peak, and he's going to be stuck. On average, this isn't going to be a very good solution.

As this example suggests, just climbing uphill is not a very good approach to finding a solution unless the problem is really simple; unless it's a Mount Fuji. Computer scientists refer to this sort of search as a greedy algorithm; as the name suggests, being greedy isn't necessarily a good thing. So what we need is a more nuanced search approach that's going to enable our hiker to get off these local peaks. This search approach has to do a little bit more exploring and little bit less exploiting than the greedy algorithm.

The search approach I'm going to describe is called simulated annealing. Simulated annealing is a search algorithm that was designed by computer scientists actually working at Los Alamos National Labs as part of the Manhattan Project. We're going to study this search algorithm for three reasons: First reason, it works; so that's a good thing, that's always a good reason to study it. Second, it's going to provide a segue into the concept of self-organization, which we're going to study in a later lecture. Third, it's going to show this sort of fan-out nature of complex systems; the idea that we're really focusing on concepts and ideas that apply widely; they fan out into a lot of disciplines. What we're doing here is we're taking a really cool idea from physics and engineering, we're turning it into an algorithm from computer science, and we're going to apply it to people trying to find a good deli. This is just great; it's totally fanning out.

A good place to start when you talk about simulated annealing is to talk about annealing. What is annealing? Annealing is a process that's used to harden glass and metals or to make crystals. Let's suppose I have a piece of glass or a bit of pre-formed metal, and you can think of these as a whole bunch of little particles—that's really what they are, so this isn't very hard to imagine—and what I want to do is I want to think about how they're formed. I want to introduce a stylized model of glass, metal, or crystal, and this is known as a spin glass. A spin glass is an enormous checkerboard. On each cell or square of the checkerboard is a little particle; and each particle has a spin that either is plus one (pointing up) or minus one (pointing down). Initially, I'm going to just randomly assign these spins. I'm going to freeze the system in that state, so some of the spins are up and some of the spins are down; the system is what we call disorganized.

Got the picture? There's a huge checkerboard; each square has a little particle on it that either points up or points down. What we'd like to do is get all of these particles to point in the same direction; then the spin glass would be organized. If the spin glass was organized, this would be tempered steel, hardened glass, or a crystal; there would be this structure with everything pointing in the same way. We don't care whether they're pointing up or pointing down, what we care about is they all point in the same direction. How do we do it? We anneal.

One thing we could do is we could go in and turn each particle; that won't work, it would take a ton of time and we'd need really, really small tweezers. What annealing does—and this is what glassmakers and metallurgists do—is take advantage of the fact the particles actually want to line up. If you think of these particles as like little [magnets], they literally want to point in the same direction. When the metal is too cold, when it's frozen, these particles are all stuck in this frustrated or disorganized state. Physicists call this "frustrated"; you can literally think of these little spins like people you know. If they're not organized, they're frustrated.

If we heat the metal, these particles become unfrozen; they're set free to do whatever they want to do, and they're going to try and align their spins with those of their neighbors. The trick would seem to be here to heat the metal so the particles are free to move around and line up with their neighbors; so if we put the heat up really high, they'd be all sorts of spinning around and they wouldn't settle down. That's not going to work; so what we need to do is heat the metal just enough so the particles get free enough to move but not so much that they can dance around, just spinning in all sorts of crazy directions. So if we heat it up just enough, each particle's going to try to match the direction of its neighbors. What's going to happen is locally in little regions, all the particles will align; so they'll either all point up or they'll all point down. The problem is going to be this: Now we're going to have local neighborhoods that possibly differ; we'll have one region where they're all pointing up, and an adjacent region where they're all pointing down. The result will be sort of like a camouflage pattern where some regions point up, some regions point down; and this boundary is disorganized or frustrated.

Once this camouflage pattern starts to form, what do we do? We think, "We're doing pretty well here, let's cool the temperature a bit." This is annealing. When you cool the temperature, you sort of lock in the region that's up and lock in the region that's down, but the particles on the boundary that are frustrated are going to keep flipping a little bit, because they don't know which way to go. There's still some heat in the system so the boundary can keep flipping, even though the centers stay fairly fixed. What this means, though, is that the boundary can move. As parts of the boundary start pointing up, then parts that were in the pointing down region are now on the boundary; so the boundary's going to move around and sort of get absorbed into one region or another.

Here's how this is going to happen: Imagine you have a small neighborhood of particles pointing up; say like just 16. There are 16 little holdouts surrounded by a bunch of particles pointing down. At some point the boundary of that region of the 16 particles is just going to start to flip, and that little region is going to shrink from 16 to 12 to 8 to 4, and eventually that whole region instead of pointing up will point down. Because these boundaries move eventually, the whole checkerboard is going to come under the same state; and once we have all the particles pointing in pretty the same direction—with the caveat that there may be still a few little bits of flipping around on some boundaries—we're going to lower the temperature a little bit more; so this is more annealing. As the temperature cools, eventually all the particles are going to stabilize in such a way that they're pointing in the same direction—in our case, that direction is down—and when they all are pointing down, we cool the temperature so far that the system freezes; that it becomes organized. We call this state organized, again, because all the particles are pointing in the exact same direction.

Annealing, then, works as follows: You heat up the glass or metal to free the particles from their surly bonds. Then you slowly lower the temperature so they can align locally; this creates neighborhoods or regions that point mostly in the same direction. You keep things hot enough so that the neighborhood boundaries keep moving. This allows the neighborhoods to align as the boundaries sort of move about. Once the neighborhoods are aligned, your particles are all organized, and what you do is you cool the whole thing off.

That's annealing. We now have an idea of how annealing works; we want to take that idea and apply it to our hiker on the rugged landscape. In a nutshell, we want to heat things up, and then slowly cool things down. It's a lot like how we approach the two-armed bandit problem: First we explore (that's by having lots and lots of heat), and then we exploit by cooling thing down. How fast we cool depends on properties of the particles, metals, and glasses. Metals and glasses are annealed at different temperature schedules.

Now let's go back to our landscape, and let's think about a "simulated" annealing algorithm, because we're not actually going to heat up our hiker. We say "simulated" because what we're going to do is use the idea of heating and cooling to find a point of high elevation on the landscapes. Here's how this is going to work: We want to think of the temperature as a proxy for the probability of making a mistake. When the temperature's high, that means the hiker's going to make a lot of mistakes. When it's low, he makes almost no mistakes. What is a mistake? A mistake for our hiker is going downhill instead of uphill.

Let's start out: Let's make the temperature really, really high. If the temperature's really, really high, it's as though our hiker doesn't know down from up; he's going to make a ton of mistakes. He's just like one of those little particles who are just sort of flipping up and down; he's going to be dancing all over the place. What he's doing is lots of exploring; he's going to just take a step, and he's not really going to know whether it's up or down, and so maybe he stays there or maybe he doesn't. Because the temperature isn't too high, he's probably a little bit more likely to go uphill than downhill. Now what we want to do is cool the temperature a little bit more. As we cool the temperature, he becomes a little bit more discerning. He still goes downhill occasionally, but just not as often. Eventually what we're going to do is cool the temperature all the way to zero, and he becomes sort of his old greedy self; he uses that greedy search algorithm where he only goes uphill and he never goes downhill.

We get what simulated annealing is: We started with lots of mistakes, we cool the temperature we make fewer mistakes, and eventually there are no mistakes. Does it work? Let's look at our two types of landscapes. Let's start with Mount Fuji. On Mount Fuji, instead of just climbing up a hill like he did before, our hiker sort of roams around a bit; he mostly goes up, but he sometimes goes down. As we cool the temperature, then he starts going down more and more rarely and he's mostly going up; so on balance what it means, he's going to be moving up the slope, he's just not going to be running up the slope like he did before. As the temperature gets even cooler, then he starts almost always going up; he takes many, many steps up for each step back, and he's pretty much going to be near the peak after a while. Eventually, though, when we drive that temperature down to zero—when we don't allow him to make any mistakes at all anymore—he's going to get right to the peak, and he's going to stay there forever. The result of this annealing is he's going to get to the peak, he's just not going to be as quick as his greedy self was; but the important part is that he does find the peak.

Now, let's put our hiker in the Appalachians. Again, we're going to start with a high temperature—he's going to explore a lot—and then we cool it a little bit; so maybe initially 55 percent of the time he goes up, and 45 percent of the time he goes down. Let's suppose he's at the top of a huge peak. It'll take many steps down in a row to get off that peak, so it's not likely he's going to do it. Even the slightest drift upward will be enough to keep him not at the peak, but at least near the top of this big mountain.

But let's suppose he's on a tiny little hill that's only two steps up; a little hillock. Now it's very likely because he allows himself to make mistakes that these mistakes will let him walk off that little hillock. Why is that? Even though he tends to go up, it's the fact that he allows himself to make several mistakes in a row because of the high temperature that he can get out of that little hill. Over time, the errors are going to allow him to escape almost any little hill, but not to escape big mountains. As the temperature starts to cool even more, the hiker's going to remain at or near the local peak he's on; and when the temperature cools all the way, he's going to become greedy, and he's going to rush exactly to that local peak and he'll stop. What's going to happen is annealing is going to give him a pretty good solution; not necessarily the optimal peak—the global peak but it's going to put our hiker on a fairly big hill, because annealing enables the hiker to do a lot more exploring than he previously did, it allows him to sort of get off on any little hillock, and at the same time the gradual cooling enables him to exploit what he's learned.

If we think over the two landscapes, Mount Fuji and the Appalachians, we see that by having this cooling temperature, you can sort of get off little hills and get to bigger hills; and you realize if you had an optimal cooling schedule—if you could figure out exactly how to anneal—you'd want to change that schedule as a function of the landscape. So on Mount Fuji, you'd want to cool that thing off as fast as possible. On more rugged landscapes, you'd want the cooling to be slower.

I want to take a brief sidebar here and talk about how evolution does this; how it balances exploration and exploitation. Let's restrict attention to species that have sexual reproduction. So what do we have? We have mutation and recombination of genes. When we have sexual reproduction, we have the crossing of genes, this recombination; and we also have these errors introduced by mutation. These are forms of exploration. This isn't like annealing when the temperature is high. By mixing the genes of two parents, the offspring can be thought of as new searches. What we get is that recombination and mutation are examples of exploration. But what's exploitation? Exploitation is just selection: Individuals that are more fit are more likely to produce. A better solution to the problem of how to make a cockroach is more likely to be in the population that gets mutated and recombined.

Let me play this out a little bit more for a second. Remember back to our Indigo Buntings in the previous lecture: If there's a lot of variation in the population of Indigo Buntings, what that means is it's almost like the temperature being high. It means when those Indigo Buntings reproduce, there are lots of things they can get; it's a high temperature. When the variation is low, that means there's not much diversity that's going to be produced, there's not much exploration that's going to go on; that's like having a low temperature. What evolution does is by changing the variation in the species, it can adjust the temperature.

This model—the idea that genes produce fitness—is sort of a gross simplification of how evolution works. We've talked about this before; but developmental pathways from DNA to fitness depend on all sorts of factors including RNA, epigenesis, and who knows what. Our understanding of the mapping from genotype to phenotype deepens daily. That said, what we're doing here is we're painting in some very broad strokes, and the crude picture has some value. What that picture shows us is that evolution is confronted with this same problem of how do we balance exploration and exploitation, and how do we do this through selection, recombination, mutation, and population variation?

We now have this basic understanding between exploration and exploitation, and we want to apply it now to dancing landscapes; to complex systems. I want to look at a particular complex system first in some detail, and then from there we'll draw some conclusions. What I want to look at are leafcutter ants. Some background: Leafcutter ants reside in tropical or semi-tropical regions. These are farmers, literally: They raise crops. They don't raise peas or beans or corn, what they raise are fungus. How do they do this? They go up and they climb trees and they cut leaves from the trees with their little mandibles; this isn't done in some sort of pell mell fashion. Leafcutter ants come in whole variety of sizes, and each one of these different sizes plays a different role. Once they've cut these leaves and carried them back to the nest, they have ants that chew them. The bigger ants first chew the leaves into smaller pieces. Then the smaller ants chew these remnants into even tinier fragments. Finally, there are these itty bitty little ants that chew these fragments into a pulp that gets consumed by a fungus known as gondylidia.

These aren't cute little colonies; they're massive productive systems, like giant factories. A leaf cutter colony can have 8 to 10 million residents. There are instances of leaf cutting colonies forming what are called "foraging lines"; they're like superhighways, 250 yards long, from a tree all the way down to this nest. We'll talk about how these paths get created when we talk about emergence in a later lecture.

For the moment, what I want to think about is we have these farmer ants. You might say what does that have to do with dancing landscapes and exploration versus exploitation? First, the trees are trying to avoid getting eaten by these ants, or at least they want to limit the damage; so we have this standard sort of co-evolutionary story going on there between the trees and the ants. But it's more complicated than that. You have these ants, then, that take these leaves and they create a fungus. The fungi the ants are producing attract bacteria, so when they build this giant mushroom factory in the forest, what the leafcutter ants are doing is they're creating a giant food source for bacteria. They're basically putting up a giant sign, "All bacteria eat here," as well as for themselves; this diner serves both themselves and the bacteria.

Here's the problem: Bacteria can reproduce a lot more quickly than ants; so how are the ants possibly going to fight the bacteria if the bacteria are able explore much more quickly than the ants do? What the ants do to combat these bacteria is they've actually evolved an antibiotic line of defense. How do they do this? The ants, it turns out, have their own bacteria; they have a bacteria that grows on their backs, literally. These bacteria that grow on their backs can attack the bacteria that eat the fungus. So what we see here in this very brief description of an ecosystem shows is this constant balance between exploration and exploitation; we have the ants trying to exploit the trees and trying exploit their knowledge to build this sort of fungus factory, and at the same time they constantly have to be exploring, using the bacteria on their back, new ways of fighting the bacteria that's eating their fungi. There are constantly reasons to explore. What happened is that leafcutter ants had to form a relationship with the bacteria in order to evolve fast enough; in order to keep exploring. Metaphorically, we can think of this as maintaining their elevation on this dancing landscape.

This logic holds writ large: What was once a very good solution could be a lousy solution once a landscape has shifted. Anyone or anything placed on a dancing landscape—whether it's a bluegill, Microsoft, or your Aunt Tessie—had better maintain some amount of exploration, just like sharks have to keep moving or risk death. The same is especially true for firms; they must, as the title of a famous management book says, innovate or die.

Now we're at sort of, I think, the key point in this lecture. We see that if the landscape is dancing, it's incredibly important to keep exploring. We saw how on a rugged landscape there's this balance between exploration and exploitation, but we can sort of anneal over time and eventually do more and more exploitation; how early on it made a lot of sense to explore, but later on we want to exploit. This was the undergirding for the algorithm simulated annealing: We start out exploring, and we end up exploiting. But on dancing landscapes—and this is the key point—the agents can never stop exploring. They can't; if they did, what they'd find is that what was once a peak might now be a valley. This, more than anything else, may explain why we see so much complexity.

Think back to our leafcutter ants and the bacteria in the trees: You have this collection of coupled landscapes. Or, let's think of a collection of competing firms. Suppose every firm is sitting at a local peak; suppose we don't have complexity, we're in some sort of nice equilibrium, and these dancing landscapes have become fixed. Once this happens—once nobody else is moving and the landscapes stop dancing—now there's incredible opportunity to explore for new and better peaks, because you have a fixed problem; we talked about this in the context of the Manhattan Project. Once you have all this time to look for a better peak, if you find it—if you find some better solution—you're going to take that action; but when you take that action, you make the landscapes for the other firms dance, so those other firms are then going to start moving around on their landscapes.

Agents that explore, whether they're firms or the bacteria on the backs of leafcutter ants, are preventing stasis; they're keeping the system from churning. Now we understand why adaptation keeps the landscapes dancing, but that begs the question: Why don't we get utter randomness? Why don't we just have the whole thing constantly in flux? That's a great question. If all the firms were constantly changing their strategies and prices, or if all the species in an ecosystem were constantly moving, then a really good strategy would just be to stick to something basic; it'd just be too crazy out there, it'd just be this mangle. If in the presence of total randomness it becomes impossible to explore and you just stick to some simple thing and exploit, you're going to cause the system to stabilize, because the benefits from exploration have fallen so much that the relative benefits of exploitation increase and it makes a lot of sense to just stand pat. This is why systems don't get too crazy, and it's the reason why we see so much complexity about us: Exploration prevents stasis, but the necessity of predictability in order to explore puts a limit on how crazy things can actually be.

Let's tie it all together. Individual agents balance this necessity to explore and exploit. As a result, they produce complexity; because if they're not exploring enough—if the system is stable—there's incredible incentive to explore. If too many people are exploring, then there's an incentive to sort of exploit. What that does is keeps the dial in between. We can think of complexity as an emergent property: no one of the agents sets out to create it, it just happens from the bottom up. In the next lecture, we're going to see how complexity is just one of the many possible emergent properties that a complex system can produce.

Emergence I—Why More Is Different
Lecture 6

Systems with many parts can produce emergent phenomena that cannot be true of the parts themselves. A pool of water can be wet, but a single water molecule cannot. Differentiated cells can combine to form a heart, a lung, or a whole person. Interconnected neurons can produce consciousness. How?

In this lecture, we discuss one of the most fascinating ideas from complex systems: emergence. Emergence refers to the spontaneous creation of order and functionality from the bottom up. If we look at the physical world, we see emergent patterns at every level, from galaxies to cells. Not only do we see structure in the physical world, we also see functionality. All of this happens without a central planner. It emerges from the bottom up. Before getting into the scientific details of emergence, I am going to describe in some detail a particular emergent phenomenon: slime molds.

Slime molds are amoebalike single-celled organisms that feed on decaying plant and vegetable matter. Slime molds only become interesting when under stress, and by stress I mean a lack of food. When this happens, an individual bit of slime will secrete an enzyme as a type of warning signal. This prompts other bits of slime to secrete the enzyme as well. The enzymes create a path, and the bits of mold begin to gather in a colony called a pseudoplasmodium. Though it has no brain or heart, the pseudoplasmodium begins to travel. At some point, the colony stops moving and begins to pile on top of itself to form a stalk. The individuals that make it to the top of the stalk release mold spores that spread by wind or rain. Although they started out as identical, the individuals now perform different tasks, a process known as breaking symmetry. This is epic emergence. Individual parts bind together to create a way to survive. As amazing as slime mold is, it is nothing compared to the human brain. Consciousness is, in many respects, the ultimate emergence.

We now turn to the science of emergence. We will talk through the various definitions and types of emergence and how they occur. There are two different distinctions between types of emergence: simple versus complex and strong versus weak. Simple emergence is a macro-level property in an equilibrium system, like the wetness created by weak hydrogen bonds holding together water molecules. Complex emergence is also a macro-level property, but it exists in a complex system not in equilibrium, like the mobility of slime molds. Strong emergence says that whatever occurs at the macro level cannot be deduced from interactions at the micro level. Weak emergence says that whatever occurs at the macro level would not be expected from interactions at the micro level.

To see how emergence arises, let's consider a model borrowed from Stephen Wolfram, called the one-dimensional cellular automaton. Three light bulbs are arranged in a triangle, each with two possible states: on or off. If a simple rule is applied to the model, blinking emerges as a property of the system. A similar process occurs in the biological world in the blinking pattern of fireflies. The fact that a rule applied locally leads to a macro-level property is what is meant by the term bottom up.

Consistency allows people to figure out what to do when new domains arise.

We now move from the blinking of fireflies to the formation of culture. In each case, we get coordinated behavior at the macro level that arises through micro-level interdependencies. We define culture as a set of shared beliefs, behaviors, and routines. Let's assume that there are 100 domains in which beliefs or actions are interdependent. For each domain, we will assume that people play a pure coordination game, in which a positive payoff occurs when people do the same thing. If we send people to interact within the domain, eventually the entire population coordinates their actions.

This model does not explain the fact that cultures have meaning and coherence. To account for this, we need to include interactions between the various domains. We need to make the landscape rugged. To do this, we will assume that people also value consistency. Consistency allows people to figure out what to do when new domains arise. Our full model then is one in which people walk around and interact. They coordinate and strive for consistency. What emerges? Something that looks a lot like culture. Cultures exhibit emergent functionalities. They allow people to know how others will behave. People know how to behave in novel situations.

Blinking fireflies and emergent cultures are interesting, but they are a long way from consciousness or emotion. For now, let's move to a model that produces an emergent functionality: robustness. Imagine a row of banks that have relationships with one another—they each have callable deposits that they can withdraw from their neighbors. Each bank has a choice: to make risky loans or safe loans. The risky loans pay more, but they may make other banks nervous enough to call back their deposits. We are going to let our banks learn whether to make risky loans or safe loans. Most will begin by making risky loans, but some will see the value of making safe loans. Those that choose to make safe loans will be breaking the symmetry of the original model. Soon, a pattern emerges: three or four risky banks, one safe bank (repeating). The emergent functionality that this pattern produces is a firewall, which makes the entire system robust. Does this mean that we should allow systems to evolve and that eventually robust outcomes will emerge from the bottom up? Not necessarily. ∎

Suggested Reading

Anderson, "More is Different."

Beinhocker, *Origin of Wealth*, chap. 6.

Holland, *Emergence*.

Miller and Page, *Complex Adaptive Systems*.

Newman, "Power Laws, Pareto Distributions, and Zipf's Law."

Questions to Consider

1. Can you think of a rule birds might use to form a flock or one fish might use to form a school?

2. Do you think the human heartbeat is strongly emergent or weakly emergent? Is it a form of simple emergence or is it complex emergence?

Emergence I—Why More Is Different
Lecture 6—Transcript

In this lecture, we're going to discuss one of the most fascinating ideas from complex systems: emergence. Emergence refers to the spontaneous creation of order and functionality from the bottom up.

If we look at the physical world, we see emergent patterns at every level. There are spiral-shaped galaxies; our own solar system with the planets orbiting the sun; and even the giant red spot on Jupiter that is really a storm that has been raging for at least 350 years and is more than three times the size of the earth; all of these we can think of as emergent phenomena. If we come back down to earth, on our own planet, we see emergent structures everywhere: Mountains and rivers take wondrous shapes; animals and plants exhibit beautiful patterns. If we take out a microscope to look even deeper, we see structure upon structure. Inside our human bodies, we see respiratory systems, nervous systems, skeletal systems. Each of these systems are comprised of small, differentiated, and structured cells that interact in such a way that they create something emergent that's more than the sum of its parts.

In sum, we see this emergent structure all about us, and we see it at every scale. It's true whether we pull out a telescope to look far into space, or if we lean over a microscope to see inside; to see what things are made of. Not only do we see structure, but what's important is that these structures have functionality: hearts beat, antelope run, rivers carry water to the sea, and so on. But here's what's amazing: All of this happens without a central planner. None of it is orchestrated from the top down; but it emerges from the bottom up.

Exploring how this structure and functionality emerge from the bottom up is perhaps the most profound and interesting question in all of science, so it should come as no surprise that this question animates scientists from a variety of fields. It's a huge question in physics, chemistry, and biology; but it's also fundamental to the study of psychology, sociology, politics, and economics. After all, how does something like culture emerge? What makes for an effective city? How do we prevent the emergence of a great depression? These are all social questions that have at their core issues related to emergence.

In this lecture, we're going to learn a little about what complex systems have to say about emergence. We'll see through some examples how patterns and structures actually do emerge; and we're going to see how functionality emerges. Before getting into some deep scientific details, I want to describe a particular emergent phenomenon; and I'm going to do this in order to drive home what emergence is. I also want to sort of establish a feeling of wonder and awe about how amazing this property is. I'm going to do this by talking about something that's actually a little bit gross: slime molds.

What are they? Slime molds are amoeba- like single-cell organisms. They feed on decaying plant and vegetable matter. You might find them in these reddish-orange blobs that you see on rotting wood in forests; or if you live in the suburbs, they're that stuff that discolors your mulch. Slime molds lead a pretty simple life: They absorb microorganisms, they secrete a little waste, and then they just sort of move on about their day; they don't write plays or watch television, and they don't ride bikes or anything like that. Slime molds are interesting, though, because when under stress—and by "stress" here I mean just a lack of a food source—when they're stressed and have nothing to eat, what an individual slime mold will do is secrete an enzyme. Think of this as a warning signal. Nearby slime molds will sense this enzyme; and if they're in similar straits, they'll sort of move in the direction of the other slime mold, and when they do they'll secrete this enzyme as well. This creates sort of an even stronger enzyme path. So you can think of this as a party that's spreading by word of mouth; except for the fact that slime molds don't have mouths, what they're doing is secreting. So it's a party spreading by secretion, if you will.

It's not hard to figure out what's going to happen: As all these little creatures start heading toward some source that's releasing the enzyme, they're releasing more of it, and you're going to get a colony that begins to form. This colony looks sort of like a globule of Jell-O; and like Jell-O, the colony consists of identical cells. There's no cell differentiation; each slime mold is the same. The colony has no heart, no brain, and (like the lion) it has no courage; it's just 100,000 or so little creatures banding together to form a bigger creature. This bigger creature has a name: It's called a pseudoplasmodium. Here's where the story gets really interesting. The pseudoplasmodium can walk. The individual cells can move, but the pseudoplasmodium moves along like a slug. This is incredible; this is emergence writ large. You have this functionality; what started out as a bunch of single identical creatures is now a superorganism that sort of walks along. In fact, early scientists thought of these pseudoplasmodia as creatures; they thought they were slugs. But they're not; they're just colonies of individual identical cells.

But it gets even better than this. At some point, this society of amoebas, this colony, stops moving if it can't find food; and the individual cells—the individual slime molds—start crawling on top of one another and create a stalk. Think of this like circus performers who create some giant human pyramid. Some of these creatures who are lucky enough are going to make it to the top of this stalk, and when they do in a last-ditch effort to survive they're going to produce some little spores; these are just seeds that spawn other slime molds. These spores are just spread by the wind or the rain; and actually if you want, you can go on YouTube or some science show and watch videos of this. These things started out as identical, but now they start forming different roles; some form the stalk and some create the spores. This is known as breaking symmetry: What started out as symmetric is no longer symmetric. This idea of breaking symmetry is going to be a core concept when we talk about emergence. The slime mold is an epic emergent: We have individual parts that are identical, and they produce something in the aggregate that's for lack of a better word just amazing; and they do this as a way to survive.

It some ways it's difficult not to be awestruck by the example of the slime molds, but really if you think about it, slime molds are nothing compared to the human brain. The brain consists of billions of neurons that are constantly reconnecting and disconnecting. They communicate through chemical and electrical pathways, and as a result the brain is capable of many of the same tricks as your basic laptop computer—it can store data, it can recall things, and so on—but even more incredibly, it produces consciousness. Consciousness is, in many respects, the ultimate emergence. It's a functionality that exists at the macro level—the level of the brain—that cannot exist in the parts. A single neuron cannot be conscious, nor can a single neuron produce meaningful cognition or can it experience emotion.

Yet a collection of neurons can write this lecture and hopefully allow other people to make sense of it.

Let's move on to the science of emergence. In the early days when complex systems were just getting started (1980s), the concept of emergence was characterized by sort of a duck test: If something cool happens at a higher level, it's emergence; so if it looks like a duck, it's a duck. This is ocular science—eyeball science—and we need to improve upon that; we need some sort of formal definitions and categorizations. In this lecture, we're going to talk through some real definitions of what emergence is, and talk about types of emergence, and we'll see how it occurs. Before we continue, let me remind you that the concept of emergence really lies at the core of complex systems. Again, by definition, complex systems consist of interacting parts, interdependent parts; they're diverse. But what fascinates scientists about complex systems is their ability—maybe in some sense even their tendency—to create something new that's essential and unpredictable at a higher level, such as we see in the slime mold, or such as we see in the brain.

Let's get on to these definitions. We're going to do this by talking about classes of emergence. I'm going to make two different distinctions between types of emergence. The first is going to be between simple and complex emergence; the second is going to be between strong and weak emergence. First, simple emergence: This is a macro-level property in an equilibrium system, like the volume of a gas, the weight of a table, the strength of a piece of steel, or the wetness of water. This may seem less exciting than the slime mold—and it is—but it may be more profound; remember, a single water molecule cannot be wet. The wetness emerges from what are relatively weak hydrogen bonds. It is for this very reason that the physicist Phillip Anderson in introducing the idea of emergence to the broader scientific community coined the phrase, "More is different." Simple emergence applies to systems in equilibrium. The slime mold was dynamic; so we refer to that as an example of complex emergence. Complex emergence is a macro-level property that exists in complex systems not in equilibrium such as our mobile pseudoplasmodium slime mold or such a flame.

Now we're going to get to the point of a current controversy. Everyone—well almost everyone—accepts that some macro-level phenomena like consciousness are emergent. The controversy stems from whether or not we can or ever will understand all emergent phenomena. There are some people out there who believe that there are some macro-level properties, like consciousness, that we're never going to be able to understand; they call these strong emergent phenomena. Strong emergence says that what occurs at the macro level cannot be deduced from the interactions at the micro level. In contrast, weak emergence says that what occurs at the macro level cannot be expected from the interactions at the micro level, but we can explain it. So the slime mold is an example of weak emergence. No way would we expect the single-cell organisms to form a slug when times get tough, and then form a flower in order to reproduce; but once we see it, we can understand how it happens. We even know (or we think we know) that those that get to be the spores—those that actually get to send out the little pods—have their descendents populate future slime molds (these are in some sense the winners), and they're not stronger, smarter, faster, or better-looking than the other ones, they're just in the right place at the right time.

Emergence exists; that's great. But how does it exist; how does it arise? That's the real puzzle; that's something that a lot of us puzzle over a great deal of the time when we sit around and think about complex systems.

To see how emergence arises, I want to consider a model borrowed from Stephen Wolfram's book, *A New Kind of Science*. This model is known as a one-dimensional cellular automata. We're going to construct a small-scale version. Imagine I have three light bulbs arranged in a triangle; I have one at the top, and one on each side. Each light bulb can be in one of two states: It can be on or off. We're going to assume that each light bulb uses the following rule: If all three light bulbs are in the same state—if they're all on or all off—they're all going to switch their state. If none of the other light bulbs is in the same state as you, you switch your state; so if you're on, and the other two are off, you're going to switch. But if one of the other two light bulbs is in the same state as you and the other one is a different state, then you're going to stay put. Those are the three rules.

Let's suppose we start with two bulbs on and one bulb off. The two that are on are each going to have one other bulb in the same state, so they're going to stay on. But the bulb that's off sees no other light bulbs in the same state, so it's going to switch on. Therefore, after the first period, we're going to have all three light bulbs on. Once we have all three on, they're all in the same state, so they all switch off; but then, since they're all off, they all go on. So what we get is a blinking set of lights. In the more formal language of dynamical systems, this is called coupled oscillation. Note that no matter how we initialize the three lights, we're going to get this blinking. If we start with all three in the same state, then the blinking is going to start right away. The other possibility is we have one of the lights in a different state than the others. But this is what we just did—what we just saw—and what happens is the lights take exactly one period, then they all get to the same state, and then they start to blink. Blinking here is an emergent phenomena; it wasn't built in, it just sort of emerged from the bottom up.

This sort of blinking occurs not only in strands of holiday lights; it also occurs in the biological world with fireflies. Fireflies are the biological world's analog to blinking light bulbs. Adjacent to my house where I live in Ann Arbor is this beautiful 30-acre prairie called Dow Prairie. On warm summer evenings you can walk through this prairie and you can see fireflies blinking in the tall grasses and way up in the treetops; it's just beautiful. The fireflies are so slow you can grab them in your hand and catch them. What's interesting about these fireflies is the flickering is just completely random; it's just like little flashbulbs going off randomly, and at any instant some of them are on and some of them are off.

Now suppose I was to move to the Philippines, or New Guinea, or to a place called Elkmont, Tennessee; if I went to a prairie there it would look very different, and here's how: In those places, the fireflies would all be blinking in unison; it would be like one giant stand of holiday lights blinking on and off as one. How does this happen? This happens by a process very similar to the one I just described with the blinking lights. This coordination emerges from the bottom-up; there's no conductor, nor is there some sort of pendulum that provides a central signal telling the fireflies when to light up and when to dim down. Complex systems scholars have thought through this and uncovered the mechanism that produces this emergent blinking. How does it work? Here's a simplified version of how the synchronization occurs.

Think of each firefly as having a little clock inside it with a spinning hand that makes a revolution every 10 seconds; so when this hand reaches the top of the clock, it triggers a mechanism so the firefly blinks. Let's start with a whole bunch of fireflies: What they're going to do is they're all going to blink at random times, and they're going to look just like the fireflies in the prairie by my house. That's it, end of story; no synchronization.

Blinking fireflies adjust their clocks to match their neighbors; so if a neighboring firefly flashes, then a firefly's clock will leap ahead to try and catch up. Why they do this is unknown; some biologists speculate that the males are trying to flash first, and that those that flash first are preferred by females. So the synching really is occurring in some sort of an arms race; they're attempting to be first. For our lecture, we're less concerned with why the synchronization occurs than in how it occurs. In a formal mathematical model, what we can say is the firefly is matching the phase of its cycle to that of its neighbor. If all the fireflies locally—this is all local—adjust to try and catch up to their neighbors, then the entire population is going to get in sync. This all happens locally; that's why emergent phenomena are called bottom up. If there were a central conductor standing in the middle of the prairie waving a baton, then we'd say the blinking is enforced from the top down.

People do the same thing, just like the fireflies; we adjust our phase and our frequency. This is how if you see crowds at athletic stadiums performing the wave or producing rhythmic clapping, what they're doing is the same as the fireflies: they're getting their cycles in sync.

I want to move away now from the blinking of fireflies to talk about the formation of culture. These two are going to seem rather far removed, but we're going to see some similarities. In each case, we're going to see coordinated behavior at the macro level that arises through micro-level interdependencies. Culture is a very difficult thing to talk about; it's tricky. It has literally hundreds of definitions. For our purposes, we're going to take a very constrained definition; we're going to say culture is a shared set of beliefs, behaviors, and routines. This includes important parts of culture such as religious and artistic expression, and it does so in a very sort of abstract way.

What I want to do is I want to think that there are, let's say, 100 domains in which actions and beliefs are interdependent; remember that's a core part of complex systems, interdependency. When two people meet, their actions are interdependent. That's also true when one person locates or captures a resource; they can either share it or hoard it. That's an interdependency. When I give lectures about cultural diversity, I often talk about where people store their ketchup, because where you store your ketchup depends on where other people in your family store their ketchup. You have again an interdependency.

A domain of belief is different from a domain of action; a domain of belief would be something like the answer to the question, "Why does the sun come up?" One belief system could be that the sun is pulled across the sky by a chariot driven by a sun god. Another belief system could be that the earth is spinning on an axis so that it only appears that the sun is coming up. There are all sorts of different belief systems that people could have, and there are all sorts of different action domains that could exist. Let's just make a list and say there are 100 of these domains; and we'll write down how each person behaves and thinks in each one of these domains. In each of these domains, we're going to play a game that game theorists call a pure coordination game. We're talked about this before. What matters in this game is that people do the same thing. If we both do the same thing, we get a positive payoff; if we don't do the same thing, we get no payoff or a negative payoff. The classic coordination game involves choosing which side of the road to drive on. If everyone drives on the left, great; if everyone drives on the right, also great; but some people drive on the left, other people drive on the right: not so great. Carnage; it's going to be ugly.

Let's imagine a world where we have all these 100 domains and we have people making idiosyncratic choices. Let's just randomly assign what people do in these. If we send people out, they'll have these incentives to coordinate with other people; sort of like the fireflies want to get in synch with people. Remember the earlier lecture when we talked about adjusting the dials when we talked about the coordination game, the game we considered involved greeting, whether we shake hands, kiss, or hug. Remember in that game some of the people were shaking hands, some were bowing, and others were hugging and kissing. Over time, the less common actions—say, kissing—

disappeared because fewer people met kissers. Eventually, if everyone keeps interacting with other people, we'd expect the whole population to coordinate on a common action—perhaps bowing—because, remember, it took a long time if we just had a network of people.

This is like the fireflies: Each person is coordinating locally and the result is emergent global coordination. The pure coordination model explains why when we visit a different country much of what the people do seems odd or different from what we do. They had to coordinate with one another and so did we; and it just randomly happened that we coordinated in different ways. That's fine; that's cool; that's great. The problem with a simple model of thinking of the world and thinking of culture as just a collection of coordination games is it doesn't explain the fact that cultures have meaning, that they have coherence; and that it's often possible to predict how someone from a particular culture will act in a given situation. For example, all else equal, an American is more likely than someone from France to try something new and risky. We are much more willing to embrace uncertainty. So that's a problem: Our model of just idiosyncratic coordination won't solve that; it's too simple.

What we need to do is introduce interactions between these various domains on which we're coordinating. This is, in a way, going to make the landscape rugged. How do we do this? The way we're going to do this is we're going to assume that people value consistency. So if I choose to hug people when I meet them, I'm also going to be probably less formal in other settings as well; so I may put my feet up on the coffee table, or I just may touch people in casual conversation, it's maybe a looser culture. This consistency across domains allows people to figure out what to do when a new domain arises. An example helps here: In cultures where authority gets more respect, like Malaysia, emails tend to be very formal. People in the United States have a more laid back, equalitarian culture, and we know this because sociologists have something called a power distance scale. The United States scores low on power distance, Malaysia scores high. As a result, I get emails from my students that say, "Hey Prof, quick question." If I were in Malaysia, those same emails would say, "Dear Distinguished Professor Page, I'm sorry to take up your valuable time but ..." We see these differences.

Consistency, perhaps, has even greater influence in belief domains. So if we have people who are very scientific in one domain—say in medical research; they believe strongly in scientific medicine, eschewing unscientific-based approaches—this creates an interaction between beliefs. People who adopt these scientific explanations in one domain are more likely to accept them in another domain. What does this mean on the ground? People who believe in science-based medicine probably believe in modern astronomy. People who believe in folk medicine or mystical spirits are probably more likely to accept that the maybe the sun is pulled by some chariot across the sky. We can formalize this desire for consistency in beliefs or in actions by assuming that in addition to coordinating with other people, individuals will sometimes choose a behavior in such a way as to be consistent with themselves.

Now we have a full model; and the full model works as follows: People walk around, they meet other people and they talk about ideas and they see actions, and they coordinate with other people because they don't want to feel uncomfortable. At the same time, they have interactions within their own choices; they strive for consistency, they want to avoid cognitive dissonance. What's going to emerge from this coordination with other people and this drive for consistency? What's going to emerge is something that looks a lot like culture. Broadly speaking, we're going to get people who are coordinated; and broadly speaking, people are going to be consistent, some more than others. But there's going to be some variation: Just like we saw with our Indigo Buntings, that there is no one Indigo Bunting, there's going to be no one French person; there's going to be variation. If we include the possibility of errors, and assume different levels of acceptance for consistent coordination, we're going to see some individuality. Cultures, then, can be thought of some sort of emergent pattern: Germans seem German; the Spanish seem Spanish; and so on.

These cultures exhibit functionalities as well; it's not just the fact that there is this notion of Frenchness, there's a functionality there. What is that functionality? Let's talk about two. First, cultures allow people to know how others will behave; not exactly mind you, but you have an idea. If I go to a baseball game with someone from Japan and the home team hits a home run, I know I shouldn't hug and kiss my guest, because I know something about Japanese culture. Second, people know themselves how to behave in novel situations. If I'm Canadian—I'm not, but suppose just for fun I'm Canadian—and I find myself in a new situation, I can just ask, "What would a Canadian do?" and once I know what a Canadian would do, that's what I'll do, because it gives me some comfort in terms of my actions.

We've talked about how culture is an emergent phenomenon; now I want to talk about the emergence of a different type of functionality: firewalls. Blinking fireflies and emergent cultures are interesting, but they're a long way from consciousness. To get there, we want to take some small steps. In the next lecture, we're going to talk about a model called the game of life, and that's going to get us closer; it's not going to get us all the way there, but it's going to get us closer. For the moment, though, I want to move just sort of halfway there and talk about a particular emergent functionality: robustness.

I'm going to do this with a model of banks. It's going to be a very simple model of banks, and it's going to work as follows: I have a bunch of banks arranged in a line, and the idea is each bank has loans out to the neighboring banks; the banks on either side. You can think of these as callable deposits of money that it can withdraw from its two neighbors. Each bank has a choice: It can make a safe loan or it can make a risky loan. Risky loans are going to pay $5.00, and safe loans pay $4.00. It seems like, "Why not make the risky loan?" The problem is the risky loan can go bad. Let's suppose it goes bad 10 percent of the time. If it goes bad 10 percent of the time that means $5.00; maybe 10 percent of the time I lose the $5.00; the expected value is $4.50; so it's still better off to make the risky loan.

I want to add one more feature to my model to make it interesting; to make it complex. Let's suppose if the risky loan goes bad, that the bank doesn't just fold, what it does is it can borrow money from its two neighbors; it can just pull its deposits from the two neighboring banks. Let's also suppose that if the neighboring bank made a safe loan, that's fine; you can just borrow the money and everything's cool. But if it made a risky loan, the bank's investors are going to get worried, and they're going to say, "Wait a minute, their risky loan failed; I want you to call our deposits from our neighbors," which will be one of the neighbors on the right. Let's assume that there's some cost to pulling these loans, pulling back these deposits from neighboring banks; and that costs $1.00 or so. If that's the case, then a bank would prefer to make a safe loan than to fail or to have its neighbor fail, because you have to borrow money from the other bank. That's the model. If you were in isolation, you'd rather make a risky loan; but because you can have these cascading failures, you'd basically like to make a safe loan.

What we're going to do is start with a whole bunch of banks; some are going to make risky loans, and some are going to make safe loans. They all have the same problem initially, but they're just going to randomly make choices. Initially, most of the banks will probably choose to make risky loans; but then some of the banks are going to realize, "Uh oh, there are too many risky loans out there, and we should avoid doing so." The fact that some banks will then choose to play it safe but others will keep taking the risk is going to break the symmetry of the initial model. Remember we talked about breaking symmetry before: That refers to the fact that things that were once identical now differ. The slime mold differentiated itself so some moved up the stalk and some became spores. This concept of broken symmetry is central to emergence. For example, the skin cells on a snake start off the same, but somehow the symmetry breaks and we get wonderful diamond-shaped patterns.

OK, back to the banks; enough about snakes. In our model we have this first period: some of them make risky loans, and some don't. Some of the banks are surrounded by other risk takers, and some are surrounded by safe banks. It's this diversity that creates diverse learning environments and allows for the emergent patterns and the emergent functionality. If a bank belongs to a long stretch of banks making risky loans, they're frequently going to find out that there's this cascading failure and they have to borrow money; they're going to learn to play it safe. Banks that are surrounded by safe banks are going to learn, "Hey, nobody fails, I might as well be risky." Suppose you have a bank—let's call it bank number 54—and it makes a risky loan. If banks 53 and 55 didn't, if they made safe loans, then that bank is fine; the cascade stops. But if 53 and 55 made risky loans, then you're going to get this cascade that spreads out; and it's going to continue in both directions until we reach a bank that makes a safe loan.

What's going to happen in this complex system is we're going to get an emergent pattern, and that pattern's going to look as follows: We're going to see stretches of banks in a row—like four or five—that make risky loans; and they're going to be surrounded by banks that make safe loans. This happens because a bank that's adjacent to five or six banks making risky loans doesn't want to make a risky loan because of the possibility of a cascade. The result is this pattern: a group of risky banks, safe banks, a group of risky banks, safe banks, and so on. This is incredible; and what we're getting here is emergent firewalls. This is an emergent functionality. These safe banks are creating firewalls that prevent the entire banking system from failing from this giant spread. No one said to these banks, "We want you to create firewalls"; it wasn't like the FDIC or some government agency came down and said, "We need you to put firewalls in here." The firewalls emerge naturally from the system, and what they do is make the system robust.

This raises a great question: Does this mean we should just allow systems to evolve, and eventually we're going to get robust outcomes? That everything wonderful is just going to emerge from the bottom up? That's great, that's a totally happy thought; but it's perhaps a bit naïve. For these firewalls to emerge, the banks had to learn. That learning took place in a fixed environment; we didn't allow the banks to change connections. In the real world of mergers and acquisitions, new financial instruments, and so on, we have no guarantee that good outcomes are going to emerge; that we're going to get the right firewalls. This is a point we're going to come back to in our final lecture.

In complex systems, the only thing we can really expect—we can't expect deficiency always—is the unexpected. What emerges in a system—whether it's slime molds, blinking lights, human cultures, or networks of banks—is very hard to predict a priori; but those patterns, structures, and functionalities that do emerge are among the most wondrous parts of our world. Among the patterns that emerge are patterns of connections; this is true in the brain with neurons, and it's true in social networks of people. Studying the emergence of network patterns is one of the hot new areas in complex systems, and it's where we're going to turn next.

Our physical life takes place in geographic space, but the worlds of commerce and ideas take place in both physical and virtual space.

In this lecture, we discuss networks and how they matter for complex systems. Over the last 20 years, network theory has burst into the mainstream. Networks, and space more generally, are central concepts within complex systems. In complex systems, space matters; the connections between people, ideas, and species influence how events play out. Models used in textbooks and taught in universities often leave space out. There are two reasons for this. Good modeling requires simplification. Networks and space were once thought to be superfluous. Networks were thought to be too difficult to model. There have been recent breakthroughs; we have new models of networks, and we have agent-based models. In some cases, networks do not matter, and in others they do, as can be seen in the spread of disease.

Complex systems models take spacing seriously. We start with some basic terminology and measures of networks. We talk about some common network structures. We discuss how those networks came to be. Finally, we discuss some properties of these various networks.

Let's start with some basics of networks. A network consists of nodes and edges. Nodes are things, and edges are relationships or connections. A network is said to be connected if you can get from one node to another. We can construct some measures of the system that tell us something about it. We can calculate the degree of each node, meaning the number of edges connected to it. We can compute the path length, or distance, between two nodes. This is the minimum number of edges that you have to move along to get from one node to another. Averages can be computed for both degrees and path lengths in a system. We can explore the properties of these measurements by using the United States as our model, where each state is a node and each border is an edge. We can compare this to a hub-and-spokes network, like those that airline and delivery companies have developed. The hub-and-spokes network is top down, but we want to talk about emergent networks. We will start with something called the random connection model. The result is something akin to the six degrees of separation experiment. These networks are fun, but let's turn to something more important: social networks.

Social networks are not random, in that your friends are also likely to be friends with each other. This is known in network theory as clustering. How, if social networks are clustered, can they also exhibit the low average path length seen in the six degrees of separation phenomenon? The solution lies in something called a small-worlds network. I have my close-knit friends, called clique friends. In addition, I have friends I have met incidentally, called random friends. A small-worlds network consists of lots of clique friends and some random friends. This network has a low average path length; the random friends extend our connections.

> **Social networks are not random, in that your friends are also likely to be friends with each other. This is known in network theory as clustering.**

What is the structure of the World Wide Web? It is neither a random network nor a small-worlds network. It is what we call a power-law network. We call it this because the distribution of the degrees of nodes satisfies a power law. Power-law distributions have lots of nodes with very low degree and a few nodes with very high degree. If we had a huge graph with a million nodes with degree 1, it would have 250,000 nodes with degree 2, 10,000 with degree 10, and 100 with degree 100. This is called long-tailed distribution.

In each of the cases we have seen, structure emerges. This raises the question: How do these structures emerge? Are they weakly emergent or strongly emergent? The answer is that they are weakly emergent. To understand why, we construct a model. We assign people various attributes and allow them to interact. People will tend to hang out with people like themselves, forming cliques. Random connections come from the fact that people have friends and relatives in different locations. This explains small-worlds networks, but power-law networks are more difficult to explain. For this, we use what is called a preferential attachment model. We use the World Wide Web as an example. When a new website appears, it wants to connect to websites with the most links. This seems to give preference to those websites that were created earliest. However, some of the most popular sites were latecomers. Therefore, the model was amended to allow sites also to have quality. Higher-quality sites exhibited attractions as well as higher-volume sites. This extended model generates an emergent long-tailed distribution.

Now we want to talk about a different property of networks (and also a general property of complex systems): robustness. We say that connectedness is the ability to get from any node to any other node. In terms of the Internet or power grids, connectedness is central. To determine the robustness of a network, we can perform knockout experiments. In a knockout experiment, we remove nodes and ask if the network remains connected. If we randomly knock out a few nodes in a power-law network, chances are that the network will remain connected. If we strategically knock out the highly connected nodes, the network will fail. Therefore, a power-law network is very susceptible to strategic attack. This does not mean that the network is in a critical state where large events are to be expected. We should see networks themselves as complex systems with diverse interacting adaptive components. ∎

Suggested Reading

Epstein, *Generative Social Science.*

Jackson, *Social and Economic Networks.*

Newman, Barabasi, and Watts, *The Structure and Dynamics of Networks.*

Watts, *Six Degrees of Separation.*

Questions to Consider

1. Try to connect yourself to the president of the United States in a path length of 6 or less. If you succeed, see if you can connect to Elvis in less than seven steps.

2. Think through whether a small-worlds network would be robust (i.e., whether it would stay connected) if you had a random or strategic knockout of nodes.

Emergence II—Network Structure and Function
Lecture 7—Transcript

In this lecture, we're going to discuss networks and their role in complex systems. Over the past 20 years or so, network theory has burst into the mainstream. Everyone became aware of electronic and virtual networks with the rise of the Internet and the World Wide Web; and the concept of a terrorist network entered the public consciousness following the atrocities of 9/11. Now, with social network sites on the World Wide Web, people create and map out their own networks. As a result, we now see networks everywhere.

Networks, and space more generally, are central concepts within complex systems. In complex systems, space matters; that how people, ideas, and species are connected influences how events play out. This is a central theme in complex systems. This may seem obvious, very obvious; yet the models written up in textbooks and taught in universities often leave space out. The standard models of supply and demand, of the spread of a disease, or of predator/prey dynamics all omit space. The models assume in effect that all of the action takes place on the head of a pin.

Why would this be, you ask? There are two reasons. The first is good modeling requires simplification; and it's the case that space and networks were thought superfluous. Second, networks were thought to be too difficult to model. But recently, we've had two breakthroughs: We have new models of networks such as the small worlds model that we'll talk about in this lecture; and then we have agent-based models which are a computational approach to studying complex systems that we're going to cover in the next lecture. In the past, people who thought that networks mattered just didn't have the tools with which to go about studying their effects. There's the old saw about a theoretician constructing a hammer and then seeing nails everywhere. In the case of networks, people saw nails everywhere they had no hammer. Now they have a hammer, and they're pounding away.

That said, these earlier models that ignore space often work pretty darn well. For example, if we want to know the price of bread, we can approximate the average price without including something like geography. In other cases, it's going to be the case that networks do matter. What I want to do is first consider a case where networks don't matter very much, and then show how if I change the model a slight bit, networks matter a great deal.

Let's consider disease spread. If we have an airborne flu virus, then a model might assume what we call random mixing; everyone on the head of a pin just randomly bumping into one another. This assumes in effect that everyone in a city bumps into every other person randomly. A classic model of disease that doesn't have networks is called the SIR model; and this is what you'd learn in a class on disease spread, an epidemiology class. In this model, people are either susceptible (S), infected (I), or recovered (R); hence SIR. The probability of a susceptible person becoming infected depends only on the percentage of infecteds in the population; that's it. The physical locations of the people don't matter, nor do friendships between the people matter.

Not all the diseases are airborne. Steven Johnson has a fascinating book called *The Ghost Map* that describes the spread of cholera in 19th-century London. Cholera spread from a contaminated well, the Broad Street well. To model that epidemic—the cholera epidemic—we'd need to take into account geography because it spread from a central source. Further, let's suppose we take a disease that spreads by needle sharing. Here we can't assume random mixing; we'd expect that the disease would spread more slowly and that it would be restricted to tight clusters of friends within some sort of network.

Complex systems models are going to take this spacing, these networks—whether it's geographical, social, or virtual—very seriously; and that's what we want do in this lecture. We want to start with some basic terminology and measures to talk about networks. That's going to give us a common language and it's going to give us a way to describe networks. Then what we're going to do is talk about some common network structures. So, for example, we'll learn that social networks don't look much like the World Wide Web. Then, we'll then talk about how those networks came to be: Why is it that social networks don't look like the World Wide Web? How did that happen? Then we're going to finish up by discussing some properties of these various networks, what we call network functionalities: How far apart are people, and whether, in fact, these networks are robust.

Notice what we're doing here. Think back to the last lecture; in that lecture we talked about emergence. Here we're going to talk about rules for connecting people to other people, or nodes in a network to other networks, and we're going to see how those rules produce network structure. These structures emerge; no one sets out to create a particular network structure. Once we know what types of structures emerge, we can then ask the question: What functionalities do those networks have? We're going to focus on two functions of networks. The first one's going to be: How quickly does information diffuse across the network? How far apart are people from one another? The second's going to be whether the networks that emerge are robust failures. When I say failure, I mean two types of failure: The first one is going to be random failure, where we just randomly wipe out a node. The second will be strategic attack, where we purposely go in and wipe out one node with the hope of wiping out the network. What we're going to see is that the Internet is robust to random failure even though that wasn't a goal of the actors who produced it; people were just making connections.

Let's start with some basics. I've been using the word "node" a bit. A network consists of nodes and edges. The way you distinguish a node from an edge is that nodes are things, like people, firms, computers, and so on. Edges represent relationships or connections between the things, the nodes. Let me repeat this because it's important: Nodes are nouns; they're people, computers, power sources, terrorists, firms, or countries. Edges are relationships between the nodes; these could be friendships, Internet connections, power lines (literally), country boundaries, or trading agreements. We're going to say a network is connected if you can get from any one node to any other node. If we took the 50 United States, they're not connected; but if we think of the lower 48 states as a network, they are.

Let's do that; let's start with the lower 48 states. We'll let the states be nodes, and we'll let edges represent shared boundaries between the states. What we have then is a connected graph: You can get from any one of the lower 48 states to any others just by passing from state to state. But again remember the entire United States is not a connected graph, because Alaska and Hawaii are not connected to any other states. Now that we have this graph or this network of the United States, we can construct some measures that tell us something about it. What are the most important measures for a network?

First, we can calculate the degree of each node. The degree of a node is just the number of edges that connect to it. Keeping with our network of the United States, let's take New Mexico. It has degree 5; it's connected to five other states: Arizona, Colorado, Utah (they just touch at the Four Corners), Texas, and—many people don't realize this either—Oklahoma; they connect at the panhandle. Therefore, we'd say that New Mexico has a degree of 5.

We can then compute the Average Degree of a network. That's just calculated by summing up the degrees of all the nodes and then just dividing by the number of nodes. It turns out the average degree of the lower 48 states— of that network—is around four. What does this mean? This means that on average, each state shares a boundary with about four other states.

Second measure: The second measure we can compute is path length or distance between two nodes. What is it? The path length between two nodes is just the minimal number of edges that you'd have to walk along in order to get from one node to the other. If two nodes share an edge, then they have a path length of 1. If they don't share an edge, but they are connected to a common node, then we'd say they have a path length of 2. The average path length, if we have a graph, is just going to equal the average of all the path lengths between all the nodes; that's a huge computation, because we have to do every single state to every other state and then take the average.

Let's have some fun; let's play a little bit with our graph of the United States. Remember, states are nodes and common boundaries are edges. The average degree of the network is approximately 4; the average path length is about 6. What does that mean? That means if we picked two states at random, we would expect the path between them to require crossing six state boundaries; those would be six edges in our graph. Let's do an example; let's take Oregon and Georgia. Let's see what the path length is. First, we can go from Oregon to Idaho (that's one); then to Wyoming (that's two); then we can cross over to Nebraska (that's three); head east to Missouri (that's four); and finally we can go to Tennessee (that's five) and south to Georgia (that's six). So from Oregon to Georgia we have a path length of 6. Some states are closer; Michigan (where I live) abuts Ohio, so the distance is 1. Others are a lot further apart; you can't get from Maine to Arizona or California in six, it would take a lot longer.

Here's what's interesting about this graph; by doing these measures we learn something: This average path length of 6 for a small graph with only 48 nodes is huge. It's especially huge for the purpose of moving between nodes; it wouldn't be very efficient. What do I mean by that? Let's suppose that we think of airlines, and let's suppose that airlines could only fly to adjacent states; so they could only fly in the edges in our graph. If you lived in Michigan, then you could fly to Wisconsin, Indiana, or Ohio. If you wanted to go to Pittsburgh, you'd have to go to Ohio first. Suppose you wanted to go from Maine to the Grand Canyon, which is in Arizona; that would take more than 10 flights, so nobody would go.

You'd think, "Wait, there has to be a way to have an average degree of around 4 or less than 4, and to have a much, much shorter average path length"; so a much simpler way to get from nodes to nodes. There is; in fact, it's possible to have an average path length of less than 2. How do you do this? You do this with what's called a hub and spokes network. Suppose you have all flights go to and from Denver; Denver, then, would be this hub. Every state is going to fly to Colorado to Denver, and then from Colorado you can fly to any other state. What we have is everything is connected to Colorado by a spoke; Colorado is this hub. The average degree here is going to be less than 2, because there are 47 states that have only 1 degree, and then there's Colorado that has a degree 47. So you have an average degree that's less than 2.

The interesting thing here is the average path length is also less than 2. Why is that? To get from any other state to Colorado, it takes only one edge; and so to get between any two states, it only takes two edges or two flights. If we want to go from Maine to Arizona, we just fly from Maine to Colorado, and then from Colorado to Arizona. Hence, it's at most two flights to get from any one state to any other.

It shouldn't be a surprise at all that airlines figured this out. So did overnight delivery companies. This explains why you always end up changing planes in Denver, Atlanta, Minneapolis, or Chicago. It's why your overnight envelopes spent the night in Tennessee. Those are the hubs in these networks. These airline networks and these package delivery networks are organized from the top down; they didn't emerge in any way. There was top down organization that put them in play.

What we want to talk about are emergent networks; we want to talk about where the networks form through a decentralized process of connecting. I want to start with something called a random connection model. I want to fill a room with 1,000 people, and I want to think of these as the nodes. Then I'm going to draw 3,000 random edges; I'm just going to randomly make 3,000 lines between people. If I do this, the resulting network is going to have an average degree of 6. Why 6? The reason why there are 6 is each of those 3,000 edges connects 2 people; we really have 6,000 connections in 1,000 people. That gives us an average degree of 6. What else might we say about this graph? Would it be connected? Given that each person is connected to, on average, six others, you can do a little math, and you can show that with very, very high probability this network is going to be connected.

What about average path length? This is a little bit trickier, but let's walk through it. Each person knows about six people; that means there are six people who are at a path length 1. Each of those 6 knows 5 other people, so that makes for 30 potential people at path length 2. There could be some overlap, so let's be conservative and say that there are 25 people at path length 2. Each of these 25 people at path length 2 has five other connections; so that means 125 people at path length 3. There could be some overlapping in here, too. What do I mean by overlap? I mean that one person that's two people away through one friend could also be two people away through another friend, and so it's the same person; we have to take into account that overlap. If we iterate one more time we're going to get 500 people at length four, and so on and so on; and what we'll see in five or six steps with this random graph we're going to have thousands of people. We're going to have the whole crowd.

This result should call to mind the famous "six degrees of separation" experiment of Stanley Milgrom; this is the idea that people are connected by six degrees. Milgrom, who is a psychologist, gave people in Nebraska a letter that was supposed to be sent to a banker in Boston. People were only allowed to send the letter to good friends; these were people who they know on a first-name basis. The letters that made it to Boston—not all of them did, but of the ones that did—the average degree (the average path length; the average number of people it went through) was about six. More recently, Duncan Watts did a much larger experiment of this form; he did it with email. He had 48,000 senders and 19 targets spread across 57 countries. The average number of edges—the average number of emails to make a connection—was also around six.

Some of you might recall a faddish parlor game called "Six Degrees of Kevin Bacon." In this game, the goal was to find a minimal path from Kevin Bacon, an actor, to another actor—let's say Glenn Close—using common movies as edges. Let's do that: Kevin was in *The River Wild* with Meryl Streep; Meryl Streep was in *Sophie's Choice* with Kevin Kline; and Kevin Kline was in *The Big Chill* with Glenn Close. So Glenn Close has a path length to Kevin Bacon of no more than 3; and this is what network theorists would call her "Bacon number," the path length between her and Kevin Bacon.

These movie star networks are fun, but they're not that important. Let's turn to something a little bit more important: social networks. Why are social networks so important? It turns out they have a big influence on the major events in our lives: who we marry, where we live, what jobs we take; those sorts of things. Social networks are not random; they don't look like that random network. The reason why is your friends are likely to be your friends. In network theory, this is known as clustering. Clustering intuitively increases path lengths. If we're all in little clusters, it's going to take a long time for me to link to someone far away. Now I have a puzzle: How can it be if social networks are clustered that they have this low average path length; the six degrees of separation phenomenon?

The solution here—and this is a brilliant solution—is something called the Small Worlds Network, and this was developed by Duncan Watts and Steven Strogatz. Let's think just for a moment about how social networks form or what they look like. I have some close-knit friends—you can think of these as my clique or my posse—and these friends are all friends with one another. If one of us slams his finger in the car door, everybody knows about it. Let's call these our clique friends. In addition, we have a few random friends—some people we went to camp with; someone we met at a wedding; maybe someone who married our sister's roommate—these people are random friends.

A small world network, in a nutshell—this is a loose approximation—consists of people who are nearby (our clique) and then some random people. If we think about it, this network actually has fairly low average path length. Why? Because we can traverse the random links to get from anyone to just about anyone else. Let's suppose I wanted to meet Kevin Bacon; I wanted to find out how close am I to Kevin Bacon? What I'd do is jump outside of my clique and I'd go to a random friend, my cousin in Los Angeles who lives a long way away. He would probably know someone in his Los Angeles clique who knows someone in Hollywood in the "industry"; so in two steps—I go to my cousin; I go to my cousin's friend in Hollywood—I'm pretty much at Kevin Bacon's door.

Social networks aren't random; they have this clustering in the random friends. What about the Internet; what about the World Wide Web? First, let me make an important distinction: The Internet is physical objects; computers, fiber optic cables, that sort of stuff. The World Wide Web is websites and the links we make connections to when we create a webpage. Let's look at the World Wide Web; what is its structure? Is it random; is it a small world? The answer is: neither; it's what's called a power law network. They call it that because the distribution of the degrees of the nodes satisfies what's called a power law (it's close to a power law). In a few lectures, we're going to spend a lot more time on power law distributions and what they mean; but for the moment, what we need to know about power laws is this: They have lots of nodes with very, very low degree, and they have a small number of nodes with really, really high degree. Most nodes aren't connected to much of anybody, and there are a few nodes that are connected to almost everybody.

Think back to our hub and spoke network. Every state was connected to Colorado, it had degree 47; the other 47 states had degree 1. That's really extreme. A power law distribution isn't quite that extreme, but it's close. In this power law distribution—here's the math—the number of nodes with degree k—like 5, 7, 9—is proportional to $(1/k)^2$. Let me say that again: it's proportional to $(1/k)^2$. So the number of nodes of degree 2 is proportional to $\frac{1}{4}$; the number of nodes of degree 3 is proportional to $\frac{1}{9}$; and the number of nodes of degree 10 is proportional to $\frac{1}{100}$. This continues all the way; so the number of nodes with degree 100 is proportional to $\frac{1}{10,000}$.

If we had a huge graph that had a million nodes with degree 1, then it would have 250,000 nodes of degree 2; 10,000 nodes of degree 10; and 100 nodes of degree 100. That's a lot of nodes with degree 100. That's why the power law distributions are sometimes called long tailed distributions, because that distribution has a long tail; there are a lot of big events. The World Wide Web is a long tailed distribution. There are a lot of people who are connected to Google, Yahoo!, and Wikipedia; but there are not many people connected to my webpage or any of my friends' webpages. I'm not alone; there are millions and millions of people like us that have very few connections. There are only a handful of sites like Wikipedia that have millions and millions of links.

Here's what's cool: The metabolic networks within species are also power laws; so are academic citation networks—who cites which papers and books—that also is a power law, it's a very long tailed distribution. There are a handful of books and papers that get cited many, many times. John Holland, who's one of the founders of complex systems, wrote a book about complex adaptive systems. This book is cited over 10,000 times; that's huge. Most books and papers get two or three citations, if that. These power law distributions—these long tailed distributions—are everywhere.

Now we have this idea of networks; we have these measures: degree, path length, etc. We know that social networks look sort of like a small world; and the Internet, World Wide Web, and citation networks have these long tails and they are sort of power law distributions. No one set out to make these structures; in each case, the network structure emerges. There's no top down organization like the airline hub and spoke network. Here's a question to ask: How does it emerge? How do we go from the micro to the macro? Can we understand it? To put this in the language of an earlier lecture on emergence, are these structures weakly emergent like the blinking fireflies and the slime mold; can we figure them out? Or, like consciousness, are they possibly strongly emergent; is it the case that we're never going to unpack these? The answer fortunately is weakly emergent; in fact, these are fairly easy to figure out.

To understand the social network, we can construct the following model. Let's assign people a set of attributes; these attributes can be things like identity characteristics: age, gender, religion. They can be hobbies: salsa dancing, Civil War reenactments, NFL football. And they can be preferences: you might like Mexican food or the Republican Party. There's a lot of research by sociologists and social psychologists that suggests that people like to hang out with people who have attributes like themselves. We'll come back to this concept in a few lectures when we talk about tipping points and the big sort; but sociologists refer to this phenomenon of hanging out with others like us as "homophily." We like to be like other people like us, and that's what our clique is. Opposites do attract; but that's the exception, really, not the rule.

So whether in a city, town, university, or a club, people are going to hang out with people who sort of look, think and act like themselves; so we get the cliques. That explains the cliques in the small world networks; where do the random connections come from? They come from the fact that we have relatives and friends who move and live in other places. We have jobs that take us all over the country or all over the world; we strike up friendships in those places as well. What's key is these friendships are all random; so the people I know in another city are different than the people that the people in my clique know in another city.

Small worlds are no mystery; it's clear why they emerge. What are a little more mysterious are these power law distributions; that's more difficult to explain. Here's one simple model that does the trick, and then we're going to amend it in just a minute. This model's called the preferential attachment model, and we're going to apply this to the Web. Suppose, though, when a website appears just randomly, it has to decide who to connect to. What it's going to do is it's going to connect to let's say four or five other sites. How does the designer choose which sites to connect to? One approach would just be to choose a site based on how many other people are attached to that site. Let me make this precise. Let's assume the probability I attach—make a link—to a given site is proportional to the size of that site. If MSNBC's site has 10 times as many links as the Boston Red Sox's site, then if I create a new website, I'm going to be 10 times more likely to link to MSNBC than I am to the Red Sox. If all website designers follow this rule—if the probability of linking to a site is proportional to the number of people who've already linked to it—then guess what? We're going to get a distribution that has a power law.

That's great; so we've explained it. Not quite: Because this process implies that sites that arrive early are going to get the most links—and that for the most part is true—and a site that enters late is only going to have a few links, so it's not likely that future sites are going to link to it. When you get this effect—this sort of first mover effect—you're going to see that sites that get on the Web early have this big first mover advantage, and these should be the successful sites. But as they're saying, while generally that's true, it's not always true.

You should see the problem: Some of the biggest sites on the Internet—Google, Wikipedia, and so on—weren't there right at the beginning. How do we explain that? To account for this, we need a better model; and Albert-Lászól Barabasi and Réka Albert have a model that amends our previous model. What it does is the following: It allows sites in the World Wide Web to also have quality; so the higher quality a site has the more people who want to connect to it. So when Google came online, even though it had very few connections, it had huge, massive, really high quality. This quality led to more links, and those links then led to even more links, until eventually Google becomes a big site. So if we take this extended model where we allow for both number of links and quality of the site to determine how many people link to a new site, we're going to get an emergent long tailed distribution. That sort of explains how the Web looks the way it does.

Now we're going to talk about network functions. If I take a complex systems approach to thinking of these networks—which is what we're doing—we think in terms of these agents (in this case the nodes) that make connections for some reason. People do it in the social world to have fun, make friends, get information, and so on; and people connect up their websites to other locations to point to interesting related topics. As we just covered, the micro level actions of these agents produce these emergent network structures; small worlds in the case of people and power law networks in the case of the World Wide Web. These emergent structures have properties; we just saw in the small worlds network that it creates this average path length that is really low. No one set out to create a low average path length, but it emerges anyway.

What we want to do now is talk about a different property of networks, and this is going to be a more general property of complex systems: robustness. Think back to the beginning to the lecture; we talked about a network being connected. A network's connected if it's possible to get from any node to any other. If we're looking at the Internet or a power grid where we have power generators as nodes and power lines as edges, if we lose connectedness then we don't have an ability for information or power to get from one place to another. This is like when your power goes out at home because a wire goes down, that's the network failing to be connected; so connectedness matters a lot.

To determine the robustness of a network, what do we do? We perform something called knockout experiments, and these are sort of fun. In a knockout experiment, we remove nodes and then we say, "As I remove these modes, does the network remain connected?" Simple experiment; but now I have a problem: which node? First, let's wipe out random nodes. Recall if we have a power law distribution, a power law network like the World Wide Web, that the majority of the nodes are only connected to a few other nodes. So I was to go in and randomly knock out 1 percent, 2 percent, or 3 percent of these nodes in a power law network, I'm not going to wipe out connectedness, because all I'm going to do is wipe out these small nodes that aren't connected to anything else. Therefore, even though no one intended for this to happen, the World Wide Web and the Internet—which also has this long tailed distribution—have emergent structures that are robust to random failures. This is great; no one set out to have this be a property, but what happens is the Internet and the World Wide Web are such that they are robust.

Wait, though; what if we consider strategic knockouts? What if we think of strategically knocking out nodes in a power law network? What would we do? We'd choose the highly connected nodes. Again, remember, in a power law network, there are a handful of very, very connected nodes. By wiping out these nodes, it's going to be relatively easy to disconnect the network. What do we have? A power law network is very robust to random knockout, and it's very susceptible to strategic attack.

In a later lecture, we're going to introduce a concept called self-organized criticality. In a system that self-organizes to a critical state, tensions are going to build up until the entire system, whether it's composed of tectonic plates or investment banks, is poised on the verge of collapse. Here we're making a distinct point, and I want to make this distinction clear. In this network case we've discussed—we're talking about agents making connections according to a rule—the resulting network, whether it's the Internet or a power grid, can be susceptible to a strategic attack. But it's not the case that the network is poised in some critical state where large events are to be expected; in fact the opposite's true: these networks are very robust from random failure. The system isn't organizing to a state where it's critical; it's organizing to a state where it's robust.

Let's wrap this up with a final thought. When we think about a network, it's natural to think of something that's fixed. But the networks we've talked about in this lecture—let's think about it; the World Wide Web, the Internet, power networks, terrorist networks, networks of banks, or even friendship networks—are in flux; they're constantly adding new nodes and edges. It's also the case that nodes and edges are disappearing. If we think back to our first lecture about what makes a system complex, one attribute was that we have diverse parts that are connected in some way over some network. As we've seen in this lecture, we can think of those structures as emergent phenomena; as being the product of the rules that the nodes follow. Most of the systems that we're going to consider in this course are constantly going to be in flux; so we should see networks not as these fixed things, but we should see networks themselves as complex systems with diverse, interacting components.

In our analysis of these networks so far, in this lecture we've talked mostly about mathematics. In the next lecture, we're going to talk about agent based modeling; this is a new methodology. That methodology is going to allow us to make these networks dance in a way; to make the networks dynamic as well.

Agent-Based Modeling—The New Tool
Lecture 8

Agent-based modeling … [has] been driving a lot of research in complex systems. Agent-based models are computer models that enable us to explore complex systems in greater detail.

In this lecture, we are going to talk about a new methodology: agent-based modeling. Agent-based models are computer models that enable us to explore complex systems. First, we talk about Philo T. Farnsworth. At the age of 14, Farnsworth thought of using lines of light on a cathode ray tube to project images sent through airwaves. In other words, he thought up the idea of television. According to legend, he got this idea from plowing lines in a field on his farm. Why are we talking about this story? To demonstrate how scientific breakthroughs such as these depend in equal parts on new ideas (the horizontal lines) and new tools (the cathode ray tube). In some cases, ideas drive the development of tools. Other times, we develop the tools first. New tools create opportunities, or they reveal information or a structure, which then results in new theories or substantiates old ones. So is the science of complex systems being driven by tools or by ideas? The answer is a little bit of both.

What is an agent-based model? Rick Riolo, a leading producer of agent-based models, describes them as consisting of entities of various types (the so-called agents) who are endowed with limited memory and cognitive ability, display interdependent behaviors, and are embedded in a network. A key assumption will be that the agents follow rules. Nowadays, the rules are written in computer code, and the behavior of the models can be watched on a computer screen. The rules that agents follow can be simple and fixed, or they can be sophisticated and adaptive.

In many agent-based models, the agents take discrete actions—they decide to move locations, switch from being cooperative to defecting, or change whether to join or exit a particular activity. Because of that, the rules they follow are threshold-based. "Threshold-based" means that the agent's behavior remains the same unless some threshold is met. Once that threshold is passed, the agent changes its behavior. These threshold effects can produce either positive or negative feedback.

Let's start with a simple agent-based model: John Horton Conway's *Game of Life*. Imagine an enormous checkerboard. Each square on this checkerboard contains an agent. That agent is in one of two states: alive or dead. Time in this model moves in discrete steps. So there is time period 0, time period 1, time period 2, and so on. In each period, each agent follows a fixed rule. The rule depends on what is happening in the eight cells surrounding the agent. Hence their behaviors are interdependent. With a simple rule, we get a blinking pattern much like we saw with the cellular automaton. What differs in this case is that the rules seem disconnected from the blinking.

Ideally, a modeling approach would have the logical consistency of equation-based models and the flexibility of verbal stories.

Agent-based models allow us to write the rules of the *Game of Life* in a computer program and to think of each cell as an individual agent. We can then start the program in a figure eight and see what emerges. What emerges, astoundingly, is a periodic orbit. That orbit happens to be of length 8. After eight periods of following the rules of the *Game of Life*, the system returns to its original configuration. In the game, it is possible to create gliders. These are configurations that reproduce themselves like the figure eight but do so one square to the left or right (or up or down). Not only does the game support gliders, it also supports glider guns. These are pulsing collections of cells that spit out gliders at regular intervals. The *Game of Life* has been proven to be capable of universal computation. Anything a computer can do, the *Game of Life* can do.

Agent-based models are also capable of what we might call high-fidelity modeling. Let's consider two such models, both dealing with potentially horrific events: fires in crowded buildings and epidemics. Imagine you are a building inspector and that you have to decide how many people can be allowed in a room. Consider two rooms, both 1,600 square feet. The first is 80 feet by 20 feet and has two doors on one of the 20-foot ends. The second room is 40 feet by 40 feet (a square) and has two doors in the middle of one side. We want to ask which room is easier to evacuate. We pull out a laptop and we write a computer program with three parts: the room, the people, and the fire. The room can be thought of as the environment, the people as the agents, and the fire as an event. In the rectangular room, our agents run toward the door, forming a line, and as a result they flow out relatively smoothly. The square room creates pileups and potential disasters. How could we prevent such disasters? We cannot make all square rooms rectangular. We can find alternatives through agent-based modeling.

One of the great concerns in modern society is the potential spread of an epidemic. How can agent-based models help us understand this process? It is possible to create an elaborate model—include every airline flight and every passenger who may or may not carry disease. We can then do experiments, such as seeing what happens if we shut down an airport. We can also understand how seemingly unimportant differences in transportation architecture could play large roles in disease spread.

All complex models can be constructed as agent-based models. What do we get out of them? What is the purpose of a model? Some people believe that the goal is to use the model to make predictions that can be empirically falsified. We can also use models to explore. We can also use models to run counterfactuals. Without agent-based models, we have two choices. We can write down a stark mathematical equation–based model, such as is used in basic economics courses or in a systems dynamics course. These are logically consistent but stark. The other alternative would be to construct a narrative, or story, of how we think events will unfold. The story has an advantage in that it is flexible. The cost of that flexibility is a potential lack of logical consistency. Ideally, a modeling approach would have the logical consistency of equation-based models and the flexibility of verbal stories. Agent-based models have both of those qualities. Constructing a complete model is often as valuable a learning experience as seeing what the model spits out. ■

Questions to Consider

1. Start with four live cells in a row, and then describe the next three periods in Conway's *Game of Life.*

2. In *Complex Adaptive Systems*, Miller and Page describe some agent-based models of a standing ovations. Try to construct your own model.

Agent-Based Modeling—The New Tool
Lecture 8—Transcript

In this lecture, we're going to talk about a new methodology: agent-based modeling. Agent-based modeling is a methodology that's been driving a lot of research in complex systems. Agent-based models are computer models that enable us to explore complex systems in greater detail. Before talking about agent-based models, first, I want to talk a little bit about Philo T. Farnsworth. Philo was born in Beaver, Utah in August, 1906, and he was by all accounts a brilliant child. At the age of 14, he hit upon the idea of using lines of light on a cathode ray tube to project images sent through the airwaves. That's right. At 14, Philo thought up the idea of television. According to legend, Philo T. Farnsworth got this idea from plowing lines in fields on his family farm, and he realized that a large picture could be decomposed into slices and then sent through the air and recombined in order to reproduce the picture. When you think about it, the television seems a little bit more like sliced bread than we might have thought.

Why this story about Philo? The television, as well as many other engineering and scientific breakthroughs—such as the germ theory of disease and the discovery of DNA—depend in equal parts on new ideas (in this case horizontal lines) and new tools (in the case of Philo, the cathode ray tube). In some cases, ideas drive the development of tools. Such was the case with the design of the atomic weapon. Scientists started with a bunch of equations, like $e = mc^2$; they knew the power lie in there, and then they developed tools to harness that power implicit in their equations. Other times what happens is we develop tools first: We get microscopes, telescopes, computers, and lasers; these are all tools that drove ideas by opening up new ways of seeing and understanding. New tools create opportunities; they reveal information or structure that results in new theories or that substantiates old ones. Take the development of x-ray crystallography, where you shoot x-rays at crystals and then analyze the scattered beams. This enables scientists to reconstruct three-dimensional arrangements of atoms and they can now understand the shape of crystals, which had previously just been conjecture.

It's also the case that new theories lead to the development of new tools. In 1964, Murray Gell-Mann and George Zweig independently theorized the existence of quarks; these are tiny subatomic particles. Quarks come in six types: up, down, top, bottom, charm, and strange. It took over 30 years until all six quarks had been found experimentally, and this used tools that hadn't even been developed when the quarks were theorized. What's interesting about this is neither Gell-Mann nor Zweig necessarily believed that quarks really existed. They constructed these things as abstract mathematical concepts, and they theorized them in order to make sense of reality; they needed their equations to work out. The empirical discovery of the quarks is an example of a theoretical idea driving science forward and then new tools being developed—in this case particle accelerators—that allowed those ideas and concepts to be verified.

In these lectures, what we've been doing is learning about complex systems; and a natural question to ask is: Is this science of complex systems—this new science—being driven by tools or is it being driven by ideas? The answer's going to turn out to be a little bit of both. So far, we've been focusing on the ideas: rugged landscapes, dancing landscapes, networks, adaptation, emergence, and so on. At the same time, though, we have to recognize that many of these ideas have been fleshed out using computer models. Up until a few decades ago, computer modeling was very costly, because computers cost a lot of money; but now, on our desktops or on our laptops, we have more computational power than even existed not that long ago.

In this lecture, what we're going to do is discuss computer modeling and some of its implications. We're going to focus on a specific type of computer modeling known as agent-based models. We're going to study some specific agent-based models and see how they work; sort of break them apart and look at their guts. We're going to learn why agent-based models differ from how we think of standard equation-based models, and why they've proved so critical to the study of complex adaptive systems. Along the way, we're going to have a little fun with one of the most famous agent-based models of all time: Conway's game of life.

Let's get started; and the obvious question to ask: What is an agent-based model? Rick Riolo, who's a leading producer of agent-based models, describes them as follows: He says they consist of entities of various types (these are the so-called agents) endowed with limited memory and cognitive ability and they are embedded in some sort of network, and the behavior of these agents are interdependent. This last condition is crucial: An agent's behavior depends on what other agents do. Often these interactions, these interdependencies, are local; agents bump into other agents around them, they're not random. We'll talk about why that's important in a few minutes. A key assumption here is that agents are going to follow rules. Nowadays, the rules are written in computer code and the behavior of the models can be watched on a computer screen. But some of the earliest agent-based models were developed back when computers were rare and costly to use. Those models were sometimes implemented with pencil and paper, or in one case that we'll talk about, the game of life, on the checkered floor of the Cambridge math department.

The rules that agents follow in these models can be simple and fixed, or they can be rather sophisticated and adaptive, sort of like the rules we use. An agent-based model of swarming bees might assume that each agent—in this case a bee—follows a very simple fixed rule. That rule might require matching the average direction of the nearest eight bees, let's say; that rule would be simple, and it's fixed. Alternatively, a rule could be adaptive. Think of a young girl deciding whether or not to go play a pick up game of soccer. She might use the rule: I'll go if the weather's temperate; so it's sunny maybe, and above 70 degrees but below 90 degrees. This seems like a perfectly reasonable rule, until we realize that if all of her friends us this same rule, too, many people are going to show up on nice days and not enough people are going to show up on bad days. As a result, our soccer player may want to adapt her rule. She may decide, "I'm going to show up on some not-so-nice days, because then I get to play more." Her rule adapts to the situation.

I need to make a quick semantic point: To change her rule, the agent was applying what we might think of as a meta rule. That meta rule was: I'm not going to get to play much soccer when it's sunny, so I should switch when I go. This meta rule is also a rule. So even though she's adapting, we can still say that her behavior is rule based.

One last point about behavior: In many agent-based models, the agents take discrete actions. They decide to move locations, they decide switch from being cooperative to defecting, or they decide to change whether or not they're going to join or exit a particular activity. Because of that, many of the rules in agent-based models can be thought of as threshold based. By that I mean that the agent's behavior remains the same until some threshold is met. Once that threshold is passed, the agent changes its behavior. These so-called threshold effects can produce either positive or negative feedbacks. We're going to examine those two types of feedbacks in great detail in the next lecture. For the moment, we want to get some grounding in how agent-based models work; so we'll hold off with deeper engagement between these positive feedbacks where you get more of the same, and negative feedbacks where you get less of the same. So that's pretty much it. Agent-based models consist of entities situated in time and space; those entities' behavior relies on rules that depend on what other agents do.

Let's start with a simple agent-based model; one of the most famous ones: Conway's game of life. This model is going to remind some of you of the simple cellular automata model that we covered in the third lecture. This model, the game of life, was developed by the mathematician John Conway of Cambridge University. The game of life works as follows: Imagine we have an enormous checkerboard. When I say enormous I don't mean enormous in the sense that each square is the size of a floor mat, I mean enormous in the sense that it has thousands of squares in every direction. Each square on this checkerboard has an agent. That agent is in one of two states: it can either be alive, or it can be dead. Time in this model moves in discrete steps; so we can think there's a time period zero, time period one, time period two, and so on. In each period, each square—or each agent—follows a simple fixed rule. That rule depends on what's happening in the eight cells surrounding the agent; so when you think of a checkerboard, if you have a cell here there are eight cells that surround it, so the behaviors are going to be interdependent.

If an agent is currently alive, it's going to stay alive if and only if either two or three of its neighbors are alive. Think of it this way: If fewer than two of your neighbors are alive, you die of boredom; if more than three of your neighbors are alive, you suffocate because there are too many people around.

What if the agent's dead? Alternatively, if the agent is dead, it can only come to life if exactly three of its neighbors are alive. How do we remember this? Think of it as two parents and a midwife. We have this very stark model here: There are dead or live agents on a checkerboard, they can't move, and they use a very simple rule.

So why the excitement? Let's start with an easy case: Suppose I have a single live agent. What happens? In the next period it dies of boredom; a short, tragic life. Let's start with two agents in a row; we have two agents next to each other. Again, they both die; because there are not enough people there, they die of boredom. Next, suppose I put three live agents in a row. The two agents on the ends are going to die because each has only one living neighbor. However, the agent in the center will remain alive because it has two living neighbors. Furthermore, the agents to the top and bottom of the center agent come to life. Why is that? It's because each one of them has three live neighbors. So the agent to the top of the center has three live agents below it, and the agent below the center has three live agents above it. So, if we put all this together, we start out with three live agents in a row, and we end up with three live agents in a column. What's going to happen next? The agent in the middle of the column is going to stay alive, but the agents on the top and bottom of the column are going to die of boredom because they have one live neighbor. But now the agents to the left and right of this center agent come back to life, and that's because each one of these is adjacent to that column and has three live neighbors.

So let's think about what happens in total: We start out with three agents in a row, we get three agents in a column. Then we have three agents in a column, and we get three agents in a row. That means we're going to get three agents in a column. What do we get? We get a blinker. This may remind some of you of the blinking lights from the earlier lecture on emergence. What differs here, though, is that these rules seem highly disconnected from the idea of blinking; blinking is really emergence. When I wrote down these rules, the last thing you would have expected as a blinker.

We've really only just begun with this game of life. Suppose that I begin with two blocks of three by three cells that are alive; so you have a block of three by three cells that are alive, and then adjacent to it at an angle I have another block of three by three cells. Think of it as a figure eight at a slight angle. What's going to happen there? We could go through the rules and figure out what's going to happen, but that would be fairly complicated; so instead what I want to do is just sort of tell you what happens. What's going to happen here is that we're going to get this elaborate pattern after the first grid and it's going to continue to move around; and then eventually, after eight periods, this thing is going to cycle back. When I start with this figure eight, I'm going to end up with a cycle of exactly eight periods where it comes back. This figure eight wasn't built in; this figure eight just sort of emerges naturally from the rules of the game of life. This is a classic example of emergence.

This may seem worthy of *Ripley's Believe It or Not*; when I start with an eight I create an eight cycle. This is spatial emergence. Nothing in the game of life suggests that it would produce a figure eight. But that's only the beginning; in the game of life, it's possible to create gliders. These are configurations that reproduce themselves like the figure eight, but they do so one square to the left or right, or up or down; they're like a walking object. If you watch a glider on the screen, you don't think of it as the rules of the game of life; you think of it as a walking object, sort of like the slime molds. Not only does the game of life support gliders—these things that move across the space—it also supports glider guns. What you get is a pulsing collection of cells that is going to spit out gliders at a regular interval.

This is going to seem like a stretch, but I'm going go back to one of the big questions of emergence we've been playing with: How does the brain, which consists of really simple parts, produce memory and consciousness? The game of life is starting with something even simpler than neurons: It just has these cells that can be on or off, and they're on a checkerboard grid; so it's not able to make new connections in any way. Yet what the game of life can do is it can create these elaborate cycles like the figure eight, and it can produce gliders that float across the screen. You can think of these regular cycles in some sense as like memory, and you're going to get these gliders as like ideas. We suddenly get this understanding that simple rules can construct things that look a lot like memory and look a lot like idea formation.

We might ask: What is the set of all possible things the game of life can do? Or another way we could ask: Is there anything the game of life can't do? From a computational perspective, the answer to that question is nothing; nada. Anything that a computer can do, the game of life can do. Anything. What do I mean by that? I mean that the game of life can count to 10; we already saw in the figure eight that it can count to 8. I also mean it can compute the square root of a number. The game of life has been proven by computer scientists to be capable of what they call universal computation. Suppose you wanted to find the cube root of 6,421. It's possible to encode that problem into a bunch of cells on a two-dimensional checkerboard that are either alive or dead; and then you can run the rules of the game of life for some number of periods; and then you can interpret what the game of life produces as an answer to that question, and it'll be the correct answer.

This may be hard to believe, I know; but it shouldn't be. Think of how a modern computer works: A modern computer consists of binary switches that go on and off. These can be programmed to do any computation you want. That's also true of the game of life. The game of life—as simple as it may be and as abstract as it may be—can do anything that a computer can do. The game of life may be too abstract; I think it appeals more to computer scientists than it to policy makers or business people. If I present the game of life to undergraduates, those who want to do mathematics find it fascinating; those who want to go be lawyers, judges, professors, or politicians maybe find it a little bit less satisfying.

But the game of life is just a warm up when we think about agent-based modeling. Agent-based models are also capable of what we call high fidelity modeling, and here's where they really link to policy. I want to consider two such models. Both are going to deal with sort of horrific events; I want to apologize for that up front. The first one's going to deal with fires in crowded buildings, and the second's going to deal with epidemics.

First, the fires: Imagine you're a building inspector and that you have to decide how many people can be allowed in a room. Suppose I have two rooms, both are 1,600 square feet; this might be a restaurant, let's say. The first one is 80 feet by 20 feet and has two doors on one of the 20-foot ends. The second room is 40 feet by 40 feet—it's a square—and it has two doors in the middle of one of the sides. I have to decide, or you have to decide, which room is easier to evacuate? Which room can I put more people in? If I'm going to appeal to aesthetics, then I would say the square room; but that's not scientific, that's just based on impression. A better approach would be to construct a model, and we could construct an agent-based model. How would we do that? It's pretty straightforward: All we'd do is we'd pull out a little laptop computer and we'd write a computer program.

That program would have three parts. In the first part, we'd define the space. We'd say, "OK, here's the room; it's a square, it's a rectangle," and we'd literally lay out the room on the computer. Part two, we'd create the people. We'd create a collection of people, and we'd place them in the room. Then what we'd have to do is we'd have to define their behavioral rule. The behavioral rule would describe how these agents are going to move in the room once the fire breaks out; what do they do? A simple rule would be something like "Run to the nearest door." Third, we'd have to create the fire; we'd have to have some way in this model of sort of creating a fire in the back of the room somewhere so then that would cause the people to run.

So if we think about this from a modeling standpoint, we can think of the room as the environment; the context. The people are going to be the agents; and the fire can be thought of as just an event. What we want to do is ask how do the agents in this environment respond to this event? We just write this model in a computer language. We could use C; we could use Java, which is an object-oriented language; or we could use a program like NetLogo, which is a software tool that's freeware developed at Northwestern University where you can construct these agent-based models very, very quickly.

Let's think through what would happen in our model. In the rectangular room, our agents are going to run toward the door, and they're going to sort of form a line, and they're just going to flow out fairly smoothly; so it's going to work quite well. What about the square room? In the square room, people head toward the door from straight on, but also from 45 degree angles, 10 degree angles, and so on; and as a result, what we'd see in this model is people are going to jam up at the door. That's going to prevent a clean exit. What the agent-based model shows us is that the square room, though it's aesthetically more appealing, isn't as good at getting people out.

The rectangular room, as long as the doors are on the shorter side, works better in terms of getting people out of the room. In the square room, you get a pile up at the door and you get potential disasters.

Now we can even go further; we could say how might we prevent these disasters? One thing we could do is we could just make square rooms rectangular; but that's going to be incredibly costly. Another thing we could do is we could say if you have a square room, you can't have as many people in there. But maybe there's a better way; maybe we can find it through agent-based modeling. One thing they did find through agent-based modeling is that if you put posts in the square room near the door, then what happens is in this frenzy as people try and escape the fire, they run into the posts; and when they run into the posts, that prevents the jam up at the door. What happens is by putting posts in front of the door, you actually enabled more people to escape the room. That's very paradoxical, but it's true.

You could say, "That's just a model and it's a simple one," but the model has been useful in helping us understand how behavior and physical space matter. Fires, though, are just small-scale disasters. They're disasters we'd like to avoid, but they're still small. We might want to ask: Can we use agent-based models to understand bigger, more complex events. The answer again here is yes. One of the great concerns in modern society is the potential spread of an epidemic. The 1918 Spanish Flu killed millions of people. With modern transportation networks, an epidemic could be much worse. A virus could board a plane from New York to Paris and be around the world in a matter of hours, literally. We might want to ask: Can agent-based models help us understand this process?

In fact, the answer's yes. It's possible to create an elaborate model—a high-fidelity model—and include every flight and every passenger. Literally: millions of passengers, and thousands of flights. If we know the nature of the disease—how it spreads, when people are symptomatic, and when they are contagious—then we can compute exact probabilities that a disease is going to jump from person A to person B and see how it spreads. We can then do experiments—we can do policy experiments—and ask: What happens if we shut down an airport? What if we quarantine a city? These are the sort of things that we never could have done with equation-based mathematical models, but we can do with agent-based computational models. We can understand also how seemingly unimportant differences in transportation architecture can play large roles in disease spread.

Take the city of Chicago; here's an interesting case. Its transportation system has what they call the loop; the loop is literally a loop of trains in the center of the city. All the trains come into the loop, they go around it, and then they branch off; this is a switching station. We might ask: Under what conditions would it be possible for a disease to stay alive on the loop itself? So the disease is literally staying alive on the loop and people are coming in, picking it up, and taking it out. To answer that question, all we'd need to know are the particulars of the disease and the particulars of how many passengers take those trains; those are both knowable. If we know how many people transferred from one line to the other on the Chicago Transit Authority, then we can figure out whether in fact this disease can stay alive on the Chicago Transit Authority. It's because all these facts are knowable that we can construct a high-fidelity agent-based model to find out whether something like a loop is a greater conduit for disease than, say, a standard hub and spoke railway system like we talked about in our network lecture. Without agent-based models, we'd have no way of answering these questions. We could pose them, but we'd have no way of addressing them in a formal way.

We've talked about three agent-based models here, and we've sort of built up in level of complexity. We started with the very simple game of life, we moved to the fire model, and now we went to the sort of high-fidelity model of epidemic. Let's think back to some of the other models that we've discussed earlier in this course: the bank failure model, the firefly model, the milk distribution model, the self-organized criticality model. All of these models really can be thought of as agent-based models. But you might ask yourself so what do we get? Why? What is the purpose of building that model?

One reason—the sort of classic reason—for building a model is to use the model to make predictions that can be empirically falsified. Karl Popper considered falsifiability to be the hallmark of science; there was a whole reason to write a model, to falsify it. Far be it for me to question Popper, let alone try to falsify him; but what I want to do is argue that in constructing agent-based models, we sort of stumble upon other reasons to model besides just falsifiability. First, we can use these models to explore. The game of life isn't producing anything that's really empirically relevant—there's no hypotheses that we can test—but it helps us get some understanding, get some leverage, as to why something as complex as cognition could emerge from simple parts. We get some idea of where memory might come from; it just lets us explore. Second, we can use these models to run counterfactuals. Here's an interesting thing to think about philosophically: We only have one world from which to get data; we only have one sort of biological history out there. We can use agent-based models to construct counterfactual worlds—we can rerun evolution—and we can understand whether an intervention might have succeeded. Suppose we're thinking about putting in a new highway or changing welfare policy: How can we understand what the implications of those choices would be unless we actually run it? Agent-based models allow us to run them.

Without agent-based models, we have two choices: We can write down a very stark closed-formed mathematical model; this is what's used in basic economics or in systems dynamics. An example of an equation might be that total demand equals a constant minus some multiple of the price. This equation would tell us how much demand for a product falls as the price rises. Equation-based models and mathematical models have one big positive: they're logically consistent. They also have one big negative: they're stark. As we try to make them more realistic, we often find that we can't solve the models. Another alternative is to construct a narrative or a story about how we think events are going to unfold. This story has an advantage because it's flexible; but we can say anything we want. Therefore the cost of that flexibility is a potential lack of logical consistency.

Let me give a very specific example. I once attended an academic talk where someone claimed using data that if a political candidate engaged in negative campaigning—if he attacked his opponent—that his share of the vote would fall. A colleague of mine said, "What if every candidate engages in negative campaigning?" The person replied that his "verbal model" implied that they would each get a lower share of the vote. My colleague said, "That doesn't make any sense, because the votes have to add up to 100 percent." The person said, "Well, you know what, that'd be a problem, wouldn't it?"

Indeed it would; and this is why verbal models aren't logically consistent. What we'd want ideally is a modeling approach that's logically consistent, but it's also flexible. Agent-based models have both those qualities. Their logical consistency is imposed by the equations put in the computer program; so any well-coded program is by definition logically consistent. But it's also true that they're flexible: Anything that you can describe, anything that you can logically write down, you can encode in an agent-based model. In fact, this is one of the great uses of agent-based models: it's just the writing down of the model. It forces you to think about all the relevant parts of the situation that you're considering. You have to ask yourself: Who are the actors; who are the players here? What is their behavior? What is the space within which they interact? How does their behavior depend on one another? When you construct a model, just the act of constructing the model is often as valuable a learning experience as seeing what the model spits out.

I want to be clear here: I don't want to say agent-based models are some sort of panacea. Yes, they're logical; and yes, they're flexible. But even if we can construct a model, that doesn't mean that we can understand what it produces. If we're confused by the world—and we often are—and we decide, "Let's construct an elaborate model to make sense of that world," we may find out that we're equally confused by our model.

Borges wrote a wonderful piece on exactly this point called "On Exactitude in Science." He tells the story about cartographers in an empire who had this incredibly ability to make maps. This ability culminated in making a map of the empire that was the size of the empire itself; it was a one-to-one mapping. That map was useless. The moral of Borges's story is that good science has to abstract from reality; it must simplify. That's how we began this lecture, talking about models simplifying. That's what good agent-based models do as well: They abstract away from the unnecessary details, and they include only what's relevant.

Learning to build a good agent-based model takes a lot of practice; it requires art and science, and both the science and the art of agent-based models are in their infancy. If you were to talk with sociologists, economists, urban planners, physicists, chemists, ecologists—all these scientists who've been using this tool; this new, exciting tool—most of them will speak as much to the potential of agent-based models as they will to its extant contributions.

An agent-based modeler at the Santa Fe Institute, which is sort of a leading think tank for complex adaptive systems, once said the following:

> At the moment, our skill at using agent-based models is akin to someone attempting to drive a car without a well-connected steering wheel. There are some people who are just stepping on the gas and careening into the guardrails on each side; and there are other people who are sort of cautiously puttering down the middle of a wide boulevard.

I think this metaphor is an apt one. The power of agent-based models has yet to be fully harnessed, but I think we're going to get there. This process is going to be spread, and sped up, by software like NetLogo that eases the burden of programming and allows the visualization of outcomes. Someday, agent-based models will be as much a part of the mainstream science as the mathematical models that we currently use in textbooks. The greater our interest in complexity, and the more the world becomes complex, the sooner that day is going to arrive.

> We're going to focus on feedbacks: positive feedbacks, in which more
> creates more; and negative feedbacks, in which more creates less.

In this lecture, we drill deeper into the implications of interdependent behaviors. We see how positive feedback creates tipping phenomena. We see how negative feedback creates stability. We see how combinations of negative and positive feedback produce path dependence. We also cover externalities. Let's get down some basic definitions. Actions that create positive feedback produce more of the same actions. In contrast, actions that create negative feedback produce less of the same actions. In complex systems models (and in the real world), agents use threshold-based rules. For an agent to act, some variable must be above or below a threshold. There is a subtle distinction between feedback, which involves interactions between instances of the same action, and externalities, which involve feedback across different actions.

In this lecture, we have four goals. First, to show how a combination of positive feedback and negative externalities produces path dependence. Second, to show how diversity produces tipping in systems with positive feedback. Third, to show how diversity produces stability in systems with negative feedback. Fourth, to show how interdependent actions can be written as a combination of feedback and externalities.

The keyboards that most of us use to type are known as QWERTY keyboards. How did the QWERTY format come to be? The QWERTY keyboard was designed to limit the likelihood of typewriter jams, and it took over the market because it created four distinct positive feedbacks. First, there was the scale of its production. Second, once you have learned to type on a QWERTY, it is easier to keep typing on one. Third, typing instruction manuals used the QWERTY design. And fourth, standardization enables resources to be shared.

The QWERTY also created negative externalities on other keyboard designs. It was not the initial state of the world that mattered for the dominance of QWERTY keyboards; it was the first few steps along the path. We refer to this as path dependence. Path-dependent processes are not predictable, a priori. The unpredictability of path-dependent processes does not stem from huge amounts of randomness; on the contrary, it depends on actions along the path.

Actions that create positive feedback produce more of the same actions. In contrast, actions that create negative feedback produce less of the same actions.

Let's now turn to our second goal—understanding how diversity plus positive feedback produce tipping. Let's model a system of 101 people fleeing a mall and see if this system can tip. We will consider two scenarios. In the first, everyone will have the same threshold, so only if the common threshold is one does one person leaving cause everyone to leave at once. If we add threshold diversity by assigning a different threshold to each person, the scenario has a tip. In complex systems, we often find that the tail (of the distribution of thresholds) wags the dog: The agents whose thresholds lie at the extremes have a large effect on the outcomes. This can also explain sorting. Thomas Schelling constructed one of the first agent-based models to explain how sorting occurs, and we can draw three lessons from Schelling's tipping model. We get segregation at the macro level even with tolerant individuals. Systems can tip; tipping goes in one direction.

We now turn to negative feedback, which tends to stabilize systems. Imagine a lake in which all of the species fix nitrogen and phosphorus levels so that the lake stays clean. Suppose we add some nitrogen to the lake. If the amount of nitrogen is small, the lake should handle it without much problem. We may be able to continue adding nitrogen and see no effect, until at some point the lake suddenly becomes eutrophic.

To see how this works, we will look at two models: an old-fashioned thermostat and a model of a beehive. The thermostat model demonstrates negative feedback because the thermostat does not perfectly stabilize the system. In the bee model, we see that genetically similar bees create fluctuations, while genetically diverse bees create stability. Additionally, in the bee model, suppose that the temperature outside gets to 120°F, and the hive temperature equals the outside temperature. We can think of this as a phase transition. The 120°F mark is a critical threshold, which is the tipping point.

To show that any interdependency can be written as a combination of feedback and externalities, let's step way back in our narrative. When we write a complex systems model, we are capturing the negative and positive feedbacks. Similarly, when we look at the effects on other actions, these either become more likely (a positive externality) or less likely (a negative externality). This sounds a lot like systems dynamics, but systems dynamics models differ from complex systems models in two important ways. Systems dynamics models do not include heterogeneity. Systems dynamics models do not include place. Does this mean that models that look at aggregates, like systems dynamics models, are not good models? Hardly. Understanding operates at many levels. When we move from aggregate-level thinking to agent-level thinking, we move from the world of systems dynamics to the world of complex systems. ■

Suggested Reading

Ball, *Critical Mass*.

Gladwell, *The Tipping Point*.

Page, "Path Dependence."

Schelling, *Micromotives and Macrobehavior*.

Questions to Consider

1. Teams must maintain an even keel. What types of feedback are necessary for stability in mood? What behaviors might produce these types of feedback?

2. Think about your own preferences about how you would like your neighbors to be. If everyone had the same preferences as you, what would residential patterns look like?

Feedbacks—Beehives, QWERTY, the Big Sort
Lecture 9—Transcript

In the previous lecture, we discussed agent-based models; we talked about how they enable us to learn about systems of interdependent agents. In this lecture, we're going to drill deeper into the implications of these interdependent behaviors, and we're going to focus on feedbacks: positive feedbacks, in which more creates more; and negative feedbacks, in which more creates less. We're going to see how positive feedbacks can create tipping phenomena in which first a few people take an action, and then a few more, until eventually everybody jumps on the bandwagon. Then we're going to see how negative feedbacks create stability; they prevent situations from getting out of hand. We're also going to see how combinations of negative and positive feedbacks produce path dependence; by this I mean that future outcomes depend on the actions along the way. Outcomes in path dependent processes can be difficult to predict and understand.

In addition to talking about feedbacks, we're also going to talk about externalities. Feedbacks refer to dependencies between the same action, and externalities refer to dependencies between different actions. Let's get these basic definitions in our heads, and then we'll start talking about the bigger issues. So positive feedbacks: Positive feedbacks are actions that produce more of the same. Think of the leaf cutter ants we talked about earlier. They left this pheromone trail to a food source; that's positive feedback. More ants, more pheromone; more pheromone, more ants; more begets more. In contrast, actions that create negative feedbacks produce less of those same actions. Rising gas prices create a negative feedback. The more gas prices go up, the less gas people buy; this in turn lowers gas prices.

In complex systems models—and in the real world for that matter—people and agents use threshold-based rules. For an agent to act—for someone to choose some behavior—a variable has to be above or below a threshold. Here's a fun scenario that I often ask my undergraduates to get them to get a deeper understanding of what threshold-based behavior is. Suppose you're in the mall and you suddenly notice that some people start running for the exits. For the sake of argument, let's suppose that in your vicinity, you can see 100 people. How many of those people would have to start running for the exits for you to join them; for you to vacate the mall? Keep in mind here you have absolutely no idea why they're leaving, they're just leaving. For most people, this number is pretty high; it's maybe 20 or 30. Whatever that number is, that's your threshold. In asking this question to students over a number of years, very few of them have thresholds above 50, and I think that's right. Think about it: If half the people in your line of sight started running out of the mall, you'd almost certainly join them; it's almost for sure that you would join the mass exodus. Why would you do this? Maybe there's a fire; maybe Elvis is in the parking lot; who knows, but people are leaving and you decide to leave. So when people run from the mall, they're creating a positive feedback, because they're increasing the probability that others run from the mall; this is just a textbook example.

A negative feedback is going to be the opposite: The more people that take action x, the less likely you're going to take that action. Let's stay right in the mall; let's not leave the mall. Suppose you walk into the mall and you're overwhelmed by the aroma of freshly baked cinnamon buns—look, I know this has happened to you; it's happened to me many times—so you make a beeline toward the cinnamon bun store. But when you get there, there's this huge line of people who had the same response you did to the smell. So here's the negative feedback: The more people in the line, the less likely you'll get in the line. This is, again, a negative feedback. At some point, the line's just going to get too long for you to wait it out; whatever that number is, that's your threshold.

Now I need to make a very subtle distinction. I need to draw a bright line between feedbacks, which involve interdependencies between the same action, and externalities, which involve interdependencies across different actions. In other words, action x creates a positive externality on action y—a different action—if the probability of taking action y increases as more people take action x. Let me give an example to get rid of all these x's and y's: The act of growing flowers creates a positive externality on making vases: the more flowers grown, the greater the need for vases. Similarly, I'll say that x creates a negative externality on y if the probability of taking action y decreases as more people take action x. Let me give an example: Waterskiing produces a negative externality on fishing, at least on the same lake. The more people that are skiing, the less fun it is to fish; and vice versa, I might add. As a lake fills with fishing boats, finding a clear alley in which to ski becomes a pretty difficult thing to do.

Now we have this distinction between feedbacks and externalities—feedbacks relate to the same action, externalities to different actions—we're ready to move forward. In this lecture, we have four goals. The first goal is going to be to show how a combination of positive feedbacks and negative externalities produce path dependence, which, as you recall, means that choices and actions along the way determine future possibilities and outcomes This is sort of a core thing within complex systems; you have all these feedbacks and externalities, you can't figure out what's going to happen. We're going to spend some time here talking about the QWERTY typewriter keyboard and how it came to be. Goal two: We want to show the importance of diversity in tipping systems that have positive feedbacks, and here what we're going to do is we're going to talk about how people sort into neighborhoods with people who look like them. Goal three: We're going to show how diversity produces stability in systems with negative feedbacks. This is sort of the main thrust here: positive feedbacks, tipping points; negative feedbacks, stability. Here we're going to talk about lake ecologies again, just for a second; and homeostatic beehives, we'll talk about how beehives maintain their temperature. Goal four: I want to convince you that interdependent actions can almost always be written as combinations of feedbacks and externalities; and this applies to anything, including the operations of, say, a farm. OK, let's go, because this is going to be fun.

Goal one: path dependence. The keyboards that most of us use are known as QWERTY keyboards; that's because of the fact that the first six letters in the top row spell out "QWERTY." You might wonder: How did QWERTY come to be? This is a great question, and a lot of research has been done on this by Paul David, an economic historian at Stanford. We might think that our old friend Frederick Taylor—he of the 21-pound shovel—figured out the optimal way to arrange keys. That's not true, and we're going to come back to that idea in just a few minutes. The QWERTY keyboard, it turns out, was designed in the 1860s by a man named C. L. Sholes. Manual typewriters worked as follows: When you pressed a key, a type bar flew up and pressed an imprint of that letter against a ribbon. If you had two adjacent type bars used frequently together, they would jam up. What a keyboard design had to do was limit the likelihood of these jams. Sholes, using data compiled by a man named Amos Densmore on letter pair frequencies, decided in looking at this data that he had to separate the letters "IE," "TH," "GH," and "CH."

This anti-jamming constraint was the reason why the QWERTY looks like it does. This still leaves open lots of options; and it's probably not a coincidence if we look at a typewriter that the word "typewriter" itself can be pecked out without lifting someone's fingers from the top row of a QWERTY. Why would this matter? This absolutely had to help typewriter salespeople as they were selling these machines door to door, because that's the first word someone would probably type out, "typewriter."

Sholes's QWERTY keyboard took over the market. Why? The reason is twofold. First: positive feedbacks. A typewriter keyboard arrangement creates four distinct positive feedbacks. Let's walk through them. First, there is what economists like to call "returns to scale in production." If you're making a whole bunch of typewriters, it's just cheaper to make them all with the same keyboard, rather than to have each one have its own idiosyncratic arrangement of the keys. Second, learning: Once somebody learns how to type on a QWERTY keyboard, they're not going to want to switch to another one. Third, teaching: If you're instructing people, there are all these manuals and courses that are developed using the QWERTY design, so it's going to be hard to switch to another one. And then fourth, sharing: If everyone has a QWERTY keyboard, then we can all use one another's typewriters, and this standardization is going to enable resources to be shared across people. We have all these positive feedbacks; it's really easy to see why QWERTYs begat QWERTYs.

But positive feedbacks aren't the only reason that QWERTYs took over; because the QWERTY also creates negative externalities. That's where QWERTYs have a negative effect on other keyboard designs. So QWERTYs not only begat QWERTYs, they wiped out everything else; they hindered other keyboard arrangements from gaining market share. You might think: Wait a minute, isn't a negative externality the same thing as a positive feedback? Not quite; let's think of the effect of QWERTYs on four-slice toasters. What's that? It's nothing; that's right: No matter how many QWERTY keyboards you buy it has no effect on toasters. But what it did do is it drove out other keyboard designs; so it had a big negative impact—a negative externality—on other keyboards. It's this lethal combination: QWERTYs begat more QWERTYs; and they also hindered non-QWERTYs. Who needs a non- QWERTY when you've already got a QWERTY? Once QWERTY got a foothold, it took over the entire market.

The key here is it's not the initial state—the first purchase—that matters; what matters here is it's the first few steps along the path. That's why we refer to things like the QWERTY keyboard takeover as path dependent. Path dependent processes are not predictable, a priori. This unpredictability stems not from huge amounts of randomness—to the contrary—it depends on sets of actions along the way; choices that are made on the path. This is what makes planning so difficult. A business might not know how many units of a product it's going to sell if consumer demand has positive feedbacks. The clothing we wear—the colors and styles—is done to match what other people wear. We buy music that other people buy. These are positive feedbacks; and what that means is we get path dependent processes and demand is difficult to predict.

Let's go to our second goal: understanding how diversity, in addition to these positive feedbacks, creates tipping; it produces larger tips. Remember by tipping, I mean that a small event or a few actions can cause a cascade and large-scale change. Let's go back to our example of people fleeing the mall and make a very explicit model. The model's going to work as follows: There are 101 people. Each one of these people has a threshold for when to leave the mall. We're going to see if this system's going to tip; we're going to see if having a few people leave the mall transforms an orderly situation into a state of panic.

I'm going to do two scenarios. In the first scenario, everybody's going to be homogeneous; they're going to have the same threshold. So the 101 people in the mall all have the exact same threshold for when to leave. I'm going to assume that everyone can see everyone else. Here's my model: Suppose there's some person in the mall—let's call her Zoey—she gets a phone call from her husband, and he's locked himself in the bathroom at home; he has a cell phone in there, so he calls her up and he says, "I'm locked in the bathroom." Zoey flees the mall to rescue her husband. What happens? What we have to do is think about those other 100 people in the mall. Each one has a threshold; what do they do? Let's suppose their threshold is 20. With a threshold of 20, when they see Zoey flee they do nothing, because they have to see 20 people fleeing before they're going to do anything. Let's make their threshold 10; again, nobody leaves. Make it 5; nobody leaves. Make it 2; nobody leaves. Only if the common threshold is 1 does someone leave, and then everybody leaves; they all leave at once, and they're just nipping at Zoey's heels, we get this mass exodus.

Let's add diversity to these thresholds and let's see what happens. We're going to do this by assigning a different threshold to every single person. Let's number the people from 1 to 100, and let's give Person J a threshold of j. This means that Person 17 has a threshold of 17; she's going to leave if she sees 17 other people leave. Person 84 has a threshold of 84; she's only going to leave when she sees 84 other people leave. Now what happens when Zoey's husband calls and she goes running out of the mall? Person 1, who's admittedly a little bit on edge, quickly follows; but no one else does. However, once Person 1 leaves the mall, Person 2 sort of freaks out and says, "Whoa, I better get out of here." Then Person 3 does, and then Person 4, and then Person 5, and so on, until eventually we get to Person 84 and she scrambles to get out of the mall as well. What we get is this tip; this cascade.

Notice in this second diverse case (and this is what's so cool about this example), the average threshold is 50—50 people—yet we still get a mass exodus because of this tip. Each person's decision to flee entices another person to flee. In the non-diverse case—in the homogeneous case—everyone could have had a threshold of 10, 5, or even 2 and there would have been no evacuation; no tip. The key point here is the average doesn't matter; what matters is what happens on the ends of this distribution, what we call the tails of the distribution. In complex systems, we often find that the tail wags the dog; the agents whose thresholds lie at the extremes have a large effect on outcomes. We've just seen in the mall example how this diversity plus positive feedbacks can produce sort of outrageous group behavior. If you think about it, how do we get celebrity, financial bubbles, and riots? The way this happens is what starts the ball rolling is not the average person, it's not the mean, it's the tail; it's those people at the extremes.

This same tail wagging the dog can explain residential sorting. One very well-established fact about American residential patterns is that we're increasingly sorted by type. We have rich neighborhoods, poor neighborhoods, white neighborhoods, and Latino neighborhoods. The economist Thomas Schelling, who is a Nobel Prize winner, constructed one of the first agent-based models, and he did this in order to explain how and why this sorting occurs. We keep talking about checkerboards because they're easy to think about; let's imagine a city as a giant checkerboard. Each square on this checkerboard isn't going to be a person, it's going to be a lot; and we're going to think of the lot as a place where people might live. We're going to put some checkers on that board, and let some of the checkers be red; and let's let these red checkers represent gun-toting, meat-eating people who hate taxes. Let's put some other checkers on that are black, and let these be tree-hugging vegetarians who like to fund large public projects. So these are people with very different political views. Let's fill in maybe half the squares.

We're going to make the following assumption about people's behavior: Each checker—which represents these sort of households—is going to look about it at the eight neighboring squares around it on the checkerboard, and it's going to look at the percentage of checkers that share its color. If that percentage is high enough, the checker is going to stay where it is; it's going to say, "Look, I'm fine; the people here are like me." If the percentage is low, it's going to move. If all of the checkers have the same threshold, and that threshold's mildly tolerant—let's suppose each one needs a third of the other checkers around it to be the same color—what we're going to see is this: Initially, there are going to be quite a few checkers that want to move, because just randomly there are going to be fewer than a third of the checkers like them.

Those relocations have two effects: They cause other checkers of the same color in their old neighborhood to want to follow suit; they want to move as well. This is a positive feedback, because once one red checker leaves, others might want to leave as well. Second, when the red checker moves to a new neighborhood, it may cause black checkers in that neighborhood to want to move. This happens because the new neighborhood has now become more red, and so the blacks may want to move out. Both of these types of movements are going to induce more movements, so the entire system can tip. If you do a careful analysis of Schelling's model, what you'll see is that the number of initial moves is about equal to the number of "tipping moves"; the moves that are caused by the initial moves. So the tipping isn't extreme, but it occurs. What's important, though, is the result of that tipping is segregation. After all of the moves, what we're left with is a checkerboard that has regions of all red and regions of all black. This is a key point, so I want to drive this home. Let's think about the agent-level behavior. The agents aren't saying, "Let's move out of the neighborhood as soon as there is one agent different from me." To the contrary. They're not even moving out when half the agents are different from them; they're only moving out if more than two-thirds of the agents are different from them. So we have at the micro level these fairly tolerant agents. What we're getting at the macro level is segregation. This is what makes Schelling's model so interesting: micro-level tolerance; macro-level segregation.

If we next introduce some diversity in the tolerance levels like we did in the mall example, the checkers that move first are the least tolerant. These least tolerant checkers are only going to be happy when they find neighborhoods that are almost entirely like them. This is going to cause the few checkers of the other color in those neighborhoods to move almost certainly; and so the result's going to be an even greater tip than we got in the original model. But this shouldn't surprise us, because we've already seen how threshold diversity combined with positive feedbacks creates these tips.

We can draw three lessons from this very simple model. Lesson number one: We can get massive segregation at the macro level even if we have tolerant individuals; so this is emergence of segregation. Lesson two: Systems can tip. A few movements can cause a mixed neighborhood to become segregated— the same could happen at a party—and this tipping becomes more likely if individuals have diverse thresholds. Lesson three (and this may be the most important for public policy considerations): Tipping goes in one direction. Once you have segregation, the systems won't tip back to the mixed state. Segregation is much more stable than the mixed configuration.

When we look at the United States, we see it's segregated in many, many ways. It's segregated by race and income, as I talked about; and evidence suggests that it's also segregated by political ideology and religion. Literally—this is true—there are Democratic neighborhoods and there are Republican neighborhoods. You might ask yourself: Is this just sort of interesting to think about, or does it matter? It absolutely matters. The more we interact with people different from us, the more comfortable we are when that interaction becomes necessary. To build up trust and understanding, we need interactions across diverse groups. Trust is extremely important for a well-grounded society; societies with higher levels of trust tend to be more successful economically and to have less crime. It's not clear how one teases out the causality—does trust drive growth and reduce crime, or does growth produce trust? Regardless, though, driving down trust levels is a policy that few people would advocate; so the segregation is something we need to be concerned about.

Now let's turn to negative feedbacks. Unlike positive feedbacks which cause a complex system to go running off in some direction, or to segregate, at a high rate of speed, negative feedbacks have the opposite effect: they stabilize systems. As we're wont to do, we're going to start really simple, and we're going to build up. Let's imagine a lake. We have a lake that contains fish, plants, and bottom sediment. Together all of these species fix the nitrogen and phosphorus levels so that the lake stays clean. No one species holds meetings to keep the lake in ecological balance, yet the balance is maintained provided the system doesn't get too much of an external shock. Suppose we add some nitrogen to the lake; that it's just runoff from our yard. If the amount of nitrogen is small, the lake should handle it without much problem. What I mean when I say "handle it" is that the algae is going to grow a little bit more, and the plant-eating species will expand their population, and so on up the food web; so the real effect on the lake is going to be negligible. In fact, we may be able to continue adding nitrogen and see no effect, until at some point all of a sudden—boom!—the lake is just going to become this slimy, algae-ridden mess; ecologists refer to this as a eutrophic lake.

To see how this works—to see how we move from sort of one state to another—what we want to do is create a model, and we're going to create a model of an old-fashioned thermostat. Then I'm going to expand this and talk about a beehive, and then I'm going to come back to the lake. First, the old-fashioned thermostat: Suppose you have an old-fashioned thermostat and you set it at 65 degrees. Let's suppose it's a warm day and the house is stabilized at 65 degrees. At 10 pm the temperature outside falls to 52 degrees and the house starts to cool down. This thermostat—which is set at 65—activates the heater, which begins to blow hot air out of the vent. When the thermostat hits 65, it's then going to shut off the heater. However, because the thermostat is a distance away from the air vent, there's still a lot of hot air blowing into the room, and the temperature in the house is going to rise eventually to say 68 degrees. As the temperature equalizes at 68, the thermostat is going to activate again and begin cooling the house; to do this, it's going to turn on the air conditioner. What's going to happen is the air conditioner is going to run until the thermostat hits 65. But unfortunately, the temperature near the vents where the cold air is coming out is now 62, so the house is going to continue to cool, and let's suppose the temperature equilibrates at 63 degrees. What's going to happen then? Guess what, the heater's going to go on. What we have is the thermostat is producing a negative feedback. When the house gets too hot, it turns on the air conditioning; when the house gets too cool, it turns on the heater. But because there are lags in the system, the thermostat doesn't perfectly stabilize the system; it causes a cycle of going up and down.

This failure of the old-fashioned thermostat begs the question: How could you stop the fluctuations? Let's look to nature, and we're going to look at bees. Beehives have to maintain a constant temperature of around 86 degrees; the reason for this is otherwise the pupa—the little offspring—are going to die. They have to be maintained at a very specific temperature. How do bees do this? It turns out bees can sense temperature. When the hive gets too cold, they're all going to huddle together like penguins at the South Pole. When the hive gets too warm, the bees are going to spread out and they're going to buzz their wings and they're going to cool things off. How do bees know it's too hot? There are no thermometers inside the hive. Bees, like humans, have these internal thermostats; and like humans, they're genetically different, so different bees feel hot at different temperatures. Two genetically identical bees are going to become hot at approximately the same temperature, but genetically diverse bees are going to become hot at different temperatures.

If we have a hive of genetically identical bees, what's going to happen is they're all going to get hot at the same time and they're all get going to get cold at the same time. We're going to have a situation just like the house with the old thermostat: The hive will get too hot, causing the bees to fan out; then it will get too cool, causing the bees to huddle over the center; then it gets too hot, and so on, and we get fluctuation. But real hives aren't genetically homogeneous, they're diverse. What if we have genetic diversity in the hive? Now the temperature at which bees get hot and cold is going to differ. If I heat up the hive, instead of all the bees getting hot at the same time, only subset of the bees are going to get hot; and as the hive starts to cool—as they start to fan it out—some of those bees, those that got hottest fastest, are going to stop, and that's going to prevent the hive from overheating. Therefore, rather than overshoot, the hive is going to stabilize at the correct temperature. You might say, "OK, this is a cute thought experiment"; but it isn't just a thought experiment, it's actually been done with real hives. They've taken hives of genetically similar bees, and what they get is fluctuation; and they've taken hives of genetically diverse bees and they've shown, in fact, that you get a nice, stable temperature.

This all seems fine and good: the pupae stay alive; the beehive is robust. But suppose the temperature outside gets really hot; suppose we crank it up to 120 degrees. At this point, it might be the case that there's nothing that the bees can do. The bees can flap like demons, but they're not going to be able to beat the heat, and the hive is going to collapse. So if we plotted the temperature of the hive as a function of the outside temperatures, we'd get a really interesting graph. As long as that outside temperature is less than 120 degrees, the hive is going to remain constant at 86 degrees; but once the temperature outside reaches 120, the hive temperature is going to jump right up to 120, because the bees can no longer make it. We're going to get what we call a phase transition. The hive has gone from a regulated phase to a hot phase; and this 120 degree mark is what we call a critical threshold. Below that threshold we're fine, above that threshold the hive collapses. The critical threshold might be referred to by some people as the tipping point.

We can finally go back to the lake: Let's return to the example of the lake that went from a clean lake—what they call an oligotrophic lake—to this eutrophic, algae-ridden lake. What we have there are negative feedbacks, just like we had in the hive. Those negative feedbacks absorb the extra phosphorous and nitrogen in the form of soil and plants; that's all good. But eventually, the systems can't absorb all of it, and just like the overworked bees the system collapses.

We can now turn to our final point. We want to show that any interdependency—any sort of dependencies between the behaviors of actors in a model—can be written as a combination of feedbacks and externalities. So I want to go way back in our narrative, back to an early lecture when we talked about writing a complex systems model. When we think about capturing the effect on others taking the same action, what we're really doing is writing down a whole bunch of negative and positive feedbacks. In our model of banks, where everybody was taking the same action—making risky loans—that created a negative feedback; as some banks made risky loans, it made risky loans less attractive. Also if we look at the effects on other actions, these are either sort of positive externalities or negative externalities.

Some of you might be thinking at this point, "Wait a minute; all these positive and negative feedbacks and negative externalities sounds like systems dynamics." It should: Systems dynamics models, like complex systems models, also include positive and negative feedbacks. That said, systems dynamics models differ markedly from complex systems models. A systems dynamics model consists of stocks and flows. We can draw these as a diagram: We put stocks in boxes, and flows are written as arrows between those boxes. A systems dynamics model of a farm would include grass and cattle as boxes; and grass would have a positive feedback on cattle, but cattle would have a negative feedback on the amount of grass.

Systems dynamics models differ from complex systems models in two very important ways. First: They don't include heterogeneity within the boxes, so all cattle are the same; they're all in the same place. We just saw the importance of heterogeneity in the bees. If we ignored that heterogeneity that would have prevented us from understanding how it is that the hive maintains this stable temperature. Second: Systems dynamics models don't include "place"; if we model that farm, we don't keep track of which grass the cattle are eating. That can matter: If we're trying to achieve an optimal balance, it requires not only having the correct quantities of grassland and cattle, it also requires rotating the cattle on the grasslands at proper intervals.

In both cases, the bees and the farm, by looking at boxes and flows we miss out on micro level detail, and that can lead us to make mistakes. Complex systems models oblige us to look at the details.

Does this mean, though, that models that look at aggregates—like systems dynamics models and some of our standard economics and physics models—aren't good models? Hardly; absolutely not true. Think back to the case of the map. Understanding operates at many levels; modeling requires simplification. If I want to understand what's going to happen in an election, I might say, "The economy is good, and this is going to have a positive effect on how voters feel about the incumbent." That's aggregate-level thinking, and it's often going to work; it's powerful and it's accurate. However, I might want to dig deeper, and I might want to ask, "How have changes in the economy affected different types of people?" When I do that—when I ask that question—I have to break open the boxes; I have to consider different types of voters; I have to put voters in different locations. I also have to model the voters as interacting with one another.

When I do these things, I move from away from aggregate-level thinking to agent-level thinking; I move from the world of sort of standard modeling or systems dynamics modeling to the world of complex systems. In this course, it's these complex systems—these complex worlds—that we're trying to make sense of. One of the things we're going to try to make sense of—we're going to talk about this in the next lecture—is how these complex systems models with their diverse agents and their positive feedbacks can produce large events. That's where we're going to turn next.

The Sand Pile—Self-Organized Criticality
Lecture 10

Complex systems often produce events whose distribution is not the traditional bell curve. One reason for that is that events are not independent; they are connected.

In this lecture, we learn the self-organized criticality theory, which explains why complex systems produce large events, like the Hatfield Airport incident. A system self-organizes if the aggregation of individual actions produces an organized pattern at the macro level. A system is said to be critical if small events trigger large cascades. Therefore, self-organized criticality implies that systems self-organize so that what emerges is critical—it can produce big events.

This lecture has four parts. First, we distinguish between normal distributions and long-tailed distributions. Second, we describe a simple random-walk model that produces a power-law distribution. Third, we discuss a model constructed by Per Bak called the sandpile model. In the final part, we will introduce a tension between complex systems thinking and optimization thinking.

First, we have normal and long-tailed distribution. Most of us are familiar with what is called the Gaussian or normal distribution: the bell curve. It is high in the middle and gradually tails off in each direction. The central limit theorem states that if we take the sum or the average of a bunch of independent random events, the result will be a bell curve. The fact that most things are normally distributed is crucial to the healthy functioning of a free and open society. In some cases, however, distribution is not normal. If we look at the distribution of sizes of wars, using deaths as a measure, we would see that most wars are very small, but every once in a while we get a huge war with many deaths. If you plot this data, you do not get a bell curve; you get a power law. Most of the events are small (this is the tall part), but huge events are possible (this is the long, flat part). Not all long-tailed distributions are power laws. Event sizes follow a power law if the probability of an event of size x is proportional to x raised to some negative power. Why and when do we not see a bell curve and instead see a power law?

Let's begin with a classic model that produces a power law distribution. Suppose we play the game called coin flipping. If we keep track of my winnings and they go 0, +1, +2, +1, +2, +1, 0, mathematicians call this a random walk. If we played the game a few million times, we could keep track of the distribution of the number of flips it took me to get back to 0. Mathematicians call these return times. The distribution of return times for a random walk is a power law. This model appears to explain the distribution of the sizes of glacial lakes.

Some systems can self-organize into critical states in which small events can trigger large cascades.

We need a model better suited to explaining wars, crashes, jams, and overruns. A candidate is Per Bak's sandpile model of self-organized criticality. Imagine a square table where individual grains of sand are sequentially dropped from above. These grains of sand accumulate until a pile begins to form. At some point, when an additional grain is added, the pile begins to collapse and some grains of sand topple from the table to the floor. If we count the grains of sand that hit the floor, we would find that most of the time only a grain or two topples the pile, but also that the avalanches are sometimes huge. It can be shown that the distribution of avalanche sizes follows a power law. Why might this and not the random walk model be a good model for explaining wars, traffic, and cost overruns?

Let's construct an even more stylized version of the sandpile model to see why. Assume the following rule: If at any time four bridge players find themselves in a square, they immediately leave the square, each one heading in a different direction. Like the sandpile model, this model self-organizes to a critical state because it starts producing cascades that follow a power-law distribution. The sandpile model and the bridge-player model produce a basic intuition that some systems can self-organize into critical states in which small events can trigger large cascades. How does this help us? First, it enables us to make better sense of the world. Second, as the world becomes more connected and more interdependent, we may be more likely to see large events. Third, the fact that we can build models may enable us to predict big events and in some cases prevent big events from occurring. Let me give three examples. First, we could stop those long tails caused by traffic jams if we could limit access to the right roads. Second, events like the Hatfield Airport incident could be prevented if connectedness is reduced (i.e., if security developed procedures to seal off portions of the airport very quickly). Finally, in the case of world wars, one good way to stop a cascade is to alleviate tension. The same goes for earthquakes. Central to our entire analysis is that the distribution of outcomes—the large events—depends on the complexity of the system. This idea that unpredictable events are not random but are the output of complexity represents a fundamental shift in perspective. ■

Suggested Reading

Anderson, *The Long Tail*.

Bak, *How Nature Works*.

Miller and Page, *Complex Adaptive Systems*.

Newman, "Power Laws, Pareto Distributions, and Zipf's Law."

Questions to Consider

1. How might a relationship between two partners self-organize into a critical state?

2. When might we want lots of small events and the rare big event and not a normal distribution of event sizes?

The Sand Pile—Self-Organized Criticality
Lecture 10—Transcript

In this lecture, we're going to talk about how complex systems can produce large events, and I want to begin by describing a story of a particular large event. On November 17, 2001, a University of Georgia football fan forgot his camera bag at the airport, so he ran back past security down the up escalator at Hatfield Airport, and disappeared into the secure region. Officials were unable to apprehend him, probably owing to the fact that there were a lot of people on a football Saturday wearing Georgia t-shirts. The entire airport had to be shut down; everyone—literally thousands of people—had to be moved outside where they waited for four hours.

The consequences of this act don't end there. Owing to the connectedness—remember that's one of our key concepts—of the airline system and the interdependencies (people who miss one flight often can't catch the next one), the repercussions of this one act extended well beyond Atlanta. A plane from Atlanta that doesn't go to Raleigh means that the same plane can't later go from Raleigh to New York. In the end, this single act resulted in the cancellation of flights all along the East Coast. Tens of thousands of people had what, for lack of a better word, was a bad travel day. This was caused by one camera bag. An event like this can happen in a complex system because the parts are connected and they're interdependent. An event like this could not have happen on the plains of Oklahoma in the 1840s. If you have a farmer who forgets his wire cutters, so he rides back to the tool shed, that isn't going to produce any repercussions that go sweeping across the plains; it ends there.

In this lecture, we're going to learn one theory of why complex systems produce these large events, like the Hatfield Airport incident. This theory is called self-organized criticality. The term is sort of self-explanatory. Recall that in an earlier lecture we discussed self organization. A system self organizes if the aggregation of individual actions produces an organized pattern at the macro level. Birds self-organize into flocks; fish self-organize into schools; pedestrians on a walkway self-organize into lanes. A system is said to be critical if small events trigger large cascades. Therefore, self-organized criticality implies what you think it would mean: It means that you have a system that self-organizes so that what emerges is a critical state; it can produce big events.

For example, one can argue that in 1914 the world political situation was critical, and that's what led to World War I. When a system is in a critical state, even a relatively minor event like the killing of the Archduke Ferdinand can set off a chain of cascading events. In that case, the minor event resulted in war throughout Europe and Asia and the deaths of millions of people. This is not good; and it's reason alone for understanding how and why criticality arises.

The potential for complex systems to produce large events doesn't mean that large events happen on a daily basis; to the contrary: Systems that self-organize into critical states produce lots and lots of small events, and what happens is this lulls us into a false sense of security and complacency. All of a sudden—boom—we get this massive catastrophe. That's what we want to understand in this lecture.

The lecture's going to have four parts. In the first part, we're going to distinguish between normal distributions and what we call long-tailed distributions. The latter—long-tailed distributions—correspond to situations in which we get lots of these small events, and then these occasional huge events. We're going to dig deeper into a particular type of long-tailed distribution called a power law, and if you remember we encountered this earlier in our lecture on networks. Second, we're going to describe a simple random walk model that gives us a power law distribution. That's going to be helpful; we're going to see how we can get a power law. But we're going to find that model to be incapable of explaining the phenomena that interest us. Third, we're going to discuss this self-organized criticality model which is called the sand pile model, and this was constructed by Per Bak. Bak calls this the sand pile model because it's a sand pile; it's literally a model of dropping grains of sand until the sand forms a pile.

This simple but evocative model is going to show us how systems can self-organize into critical states. We're going to see how this sand pile can create avalanches that have this long-tailed distribution; this power law. In this third part of the lecture, we're also going to talk about how real systems with real people—not just grains of sand—can do the same thing. Then in the final part, we're going to introduce a tension between complex systems thinking and optimization thinking. We're going to see how an obsession with optimization and efficiency can produce criticality; thus sometimes we're going to be better off building a little slack into our systems so that we don't have these large events. This is a topic we're going to come back to in a later lecture. In this lecture, we're just going to talk about this in a very confined space; we're going to talk about it in the context of supply chain management.

Let's get started. Part one: normal versus long-tailed distributions. Most of us are familiar with what are called Gaussian or normal distributions. These are bell curves: high in the middle, and gradually tails off in each direction. One of the most well-known and powerful theorems in all of mathematics—the central limit theorem—states the following: If we take the sum, or the average, of a whole bunch of independent random events, then the result will be a bell curve. For example, if I were to go to the grocery store and pull off 500 loaves of wheatberry bread, and I were to measure them—weigh them—down to the hundredth of an ounce, I would find that most of them do not weigh exactly 24 ounces as claimed on the package. Instead, some are going to weigh a little bit more; some are going to weigh a little bit less. Perhaps the yeast was more active in one loaf than another; perhaps one loaf got more or fewer wheat berries; perhaps one just got more dough.

So long as these "random events"—the wheat berries, the yeast, the dough—are independent, having more dough has nothing to do with having more yeast. As long as there's no interaction between the effects, the effect of the yeast doesn't depend (at least not that much) on the number of berries or the amount of dough, then—this is what the central limit theorem tells us—if I plot the weights of these 500 loaves of bread, I'm going to get a very nice bell-shaped curve. Most of the loaves will be within a few tenths of an ounce of the packaged weight, and it'll sort of fall off gradually. Or, for example, suppose I looked at the number of people at the grocery store on a given day, the amount of money withdrawn from a local bank, or even the size of the tomatoes in your backyard; most of these things will be normally distributed. The fact that this is true is crucial to the healthy functioning of a free and open society. I'm not kidding; this is one of the most important reasons why society works.

Suppose that the central limit theorem weren't true; just suppose for a second that it did not hold. Suppose that, instead, if you added up a bunch of random independent events, that most of the time the bread would weigh really close to 24 ounces, but that every once in a while it would come in at 40 pounds. Or suppose that the number of people at the grocery store was almost always 200 or 201, but then every once in a blue moon not 600 or 800 people showed up, but 10,000 people just dropped in. How would the store owners possibly respond to that; how would they know what to order? How could we do things like design roads and airports if instead of getting regular numbers of people, every once in a while we got huge numbers of people? Fortunately, that doesn't happen, so society works fine.

Except for the fact that in some cases it does happen. If we look at the distribution of the sizes of wars, using deaths as a measure, we see that most wars are very small. The smallest war may be the Pig War between the United States and Britain. This was over a boundary dispute near Vancouver. The history books sometimes call this the San Juan Border Dispute, but it's more popularly known as the Pig War. No men were killed, though one nameless pig did meet its death. If we look at wars and confrontations with small death tolls, like the Pig War, we're going to see they're the rule, not the exception. Every once in a while, though, we get a huge war like World War I or World War II in which tens of millions of people died. If you plot the size of war deaths, you don't get a nice bell curve, what you get is a power law. If you plot a power law—we talked about this before—you get a graph that looks like a children's slide that continues for a long, long time after it's flattened out at the bottom. Most of the events are going to be small (this is the tall part of the slide), but there are huge events that are possible (this is that long, long flat part of the slide).

Where else do we see power laws? Cost overruns on public projects have this same power law distribution. In Boston, there was a project called the Big Dig that involved building a two-mile-long underground tunnel; a road, basically. The bids for this originally were 4 billion dollars; the cost came in at 14.6 billion dollars. It's also true if we look at the size of natural events like earthquakes and floods that we're going to see power laws. I want to be clear here: Not all long-tailed distributions are power laws. If we look at the stock market, we're going to find that daily gains and losses aren't bell-shaped, and they have a bit of a long tail; but the thing is in this case, it's not quite long enough to be a power law.

To talk about a power law, I need to formalize what I mean by large events. Event sizes are going to follow a power law if the probability of an event of size x—say size 10—is proportional to x raised to some negative power; like 10^2 and then take the inverse of it. Recall from our earlier lecture on power laws in networks that if the exponent were 2, then the probability of having an event of size 100 is just going to equal $1/100^2$; that's not a big possibility, but the point is, it's still a possibility. Big events can occur, but they're rare.

To see the difference between a power law and a normal distribution—a bell curve—in a formal way, it's helpful to make a comparison. So imagine we take human heights, and we assume that they're distributed according to a power law with an exponent of one; and we're going to set the mean of this power law distribution as the same as the current existing height distribution, so it's going to have a mean of 5 feet 9 inches. If we take the bell curve—the real distribution—what we're going to get is that pretty much everybody is between 3 feet 6 inches tall and 8 feet tall; there's never been anyone over 9 feet tall.

But what if instead we had the distribution of heights satisfy a power law, but again we keep the mean at 5 feet 9 inches? Then we're going to have 60,000 people who are 9 feet tall; we're going to have 10,000 people 17 feet tall; and we'd have 1 person over 1,000 feet tall. Let's just think about how much it would cost to feed and clothe this guy. Incidentally, we'd also have 170 million people who are less than 7 inches tall. We need all those tiny people to balance out the really tall ones in order to keep the mean at 5 feet 9 inches. Luckily, people aren't power law distributed, so we never even bump into anyone even 25 feet tall. But we do get massive cost overruns; we do have large wars; and we do get massive earthquakes and floods. The question we want to ask is: Why? Why and when do we not see a bell curve—when won't we see this nice, normal distribution—and when instead are we going to see a power law, or some other long-tailed distribution?

That's not going to be an easy question to answer. So we're going to ask a simpler question; we're going to sort of do things backwards. We're going to ask: Can I come up with some models that produce power laws? Once I have a set of candidate models, I can ask whether those models fit reality. I want to admit, this is sort of backward engineering; it isn't the normal way that science is done. Normally what we do is we think about how a situation works, construct a model, and see what it produces. But for purposes of explanation this is going to be a better route; we're going to take some known models and see how they work.

I want to begin with a classic model that produces a power law; this is the first one that you'd probably learn if you studied power law distributions. This is going to come from a game—it's a very boring game, in a way—that just involves coin flipping. The rules of this game are going to be as follows: We're just going to flip a coin. Every time the coin comes up heads, you pay me a dollar. Every time the coin comes up tails, I'm going to pay you a dollar. We going to start out even Steven; the score is just zero to zero. A sequence might go as follows: heads, heads, tails, heads, tails, tails. If we kept track of my winnings, they'd sort of go as follows: We'd start out with nothing; we'd get the heads, I'm up one; another heads, I'm up two; then a tails, I'm back up to up one; a heads, I'm up two; and then tails, tails, I'm back down to zero. So if I plot this zero, one, two, one, two, one, zero, mathematicians call this a random walk; because whether it goes up or down is random, it's determined by a coin flip.

We can count how long it takes that random walk to get back to zero. In the example I just gave you, it was six flips. If we played this game for a few million rounds—if we sat around for several hundred years and just kept playing this—we could keep track of the distribution of the length of time it took me (the number of flips it took) for me to get back to zero; so when I'm at a zero, how long does it take for me to get back to a zero? Mathematicians refer to this as return time; so what we can do is look at this return time distribution. What we'd see is that most of the time it wasn't very many flips—so if I was just at zero, usually it would only take me a few flips for me to get back—but if I start winning a whole bunch of times in a row (or if you start winning a whole bunch of times in a row) it could take a long, long time for me to get back to zero. In fact, if you plot the distribution of return times for this random walk, it's a power law. So empirically that's true; it's also theoretically been show to be true. You can do a lot of math, and you can prove this would be a power law.

Great; this is cool. We have a process that generates a power law; we see these power laws in nature. Remember, again, this power law is a distribution that has many, many small events and these occasional big ones. That's what we saw in this random walk: Most of the time I come back quickly; sometimes it takes a long time. So we have a mathematical model, but is it useful? Does this explain war deaths, earthquakes, or cost overruns? We can ponder it for a while, but it doesn't seem to work. In fact, if we look across sort of all the empirical cases where we see power laws out there in the real world, this model doesn't seem to work, except for in one case; an interesting case: the distribution of the sizes of glacial lakes.

Most glacial lakes are really small—they're like ponds—but occasionally we get huge lakes. How can this random walk model explain that? Let's think for a second about how a glacier carves out a lake. Let's imagine a glacier just sort of moving across the land, like a random walk. Each time it digs down into the earth and then comes up to the surface it forms a lake. But that's just like the return times in our coin flipping model; you win for a while so I go negative, and then I keep winning, and when I get back to zero we formed a lake. If the glacier's following a random walk, the size of the lakes should be a power law. In fact, they are. That's great; our simple model gives us an explanation of why lakes are distributed the way they are.

But we need a model better if we want to explain wars, traffic jams, stock market crashes, or cost overruns. I'm going to present a model that does that. That model is Per Bak's sand pile model of self-organized criticality. Here's how Bak's sand pile model works. Imagine that I have a big square table in the center of a room; and imagine that I have individual grains of sand that are just sequentially dropped from above the table. One by one, these grains of sand accumulate until eventually we're just going to get a pile that forms. At some point, this pile is going to become large enough that when an additional grain of sand is added, the pile begins to collapse and we get an avalanche with grains of sand toppling on the floor. There's going to be an angle there that is known as the angle of repose, which is the angle that we have for this sand pile.

Suppose that we count the grains of sand that hit the floor; so when I drop one, I keep track of how many hit the floor. What are we going to find? We're going to find that most of the time nothing happens; and when we actually get something to fall, we're going to get maybe one grain or two grains that topple off the pile and onto the floor. But then every once in a while we're going to get huge, enormous avalanches; and it can be shown mathematically that if we do this, the distribution of the avalanche sizes is going to satisfy—you guessed it—a power law. The state—remember in self-organized criticality we're talking about a state—in this case it's the pile; and that's a critical state because it's poised in a configuration (a sand pile) such that additional grains of sand can possibly cause huge cascades. Most of the time they don't, but they can; that's the point: they can.

Very quick caveat: Bak's model was idealized and run on a computer; so he didn't use real grains of sand. But this model created so much excitement that efforts were made to sort of replicate this computer model with real sand. This was hard to do; it turns out real sand is too wet, so if you make a huge pile of real sand, it all gets compacted. What they found, though, is that rice does work; so if you do this experiment with real grains of rice, you get exactly what Bak showed. However, we don't call this the "rice pile model" because it's just not very evocative as an image.

So let's go back to the model. Why might this model, as opposed to the random walk model, be good at explaining things like wars, traffic jams, and cost overruns? To show you why, I want to construct an even more stylized version of the sand pile model. One thing about the sand pile model is that you get this power law distribution even if you change the assumptions in a whole bunch of ways. In this version, which is a variant of it, we're also going to get a power law; but I want you to think of it in the following way: Let's replace the table with, once again, a checkerboard; we're doing a lot of checkerboard models because they're easy to think about. When a grain of sand falls, it's going to land in one of these spaces. But now instead of thinking of grains of sand, OI want to replace these with little tiny people. Not just any people, but members of a group called Bridge Players Anonymous.

Members of Bridge Players Anonymous want to avoid meeting in groups of four, and the reason why is if four of them happen to be in the same square at the same time, a Bridge game might break out; and if they all sit there, they might go on a weekend Bridge bender and just ignore their families and ignore their work. They're members of Bridge Players Anonymous; they don't want bridge games to break out.

Let's assume the following rule: If at any time four people find themselves in a square, they leave the square; one goes north, one goes south, one goes east, and one goes west. Let's think through this model. We're going to start dropping bridge players; so squares are going to begin to get two, three players each. Eventually, a fourth player is going to land in a square. Once there are four players in one square, they all exit; they head to the four corners of the earth. Once that happens—when one of them goes to the north or south—if there are only one or two people there, then the process is going to stop. But suppose the square to the north and the square to the east both have three players. After this original exodus from the square, those two neighboring squares now have four. The four players in each of those squares scurry off; but now the square to the north east is going to get two new players, one from the north and one from the east. So if it started with only two players, the cascade is going to continue.

If we run this bridge model, what we're going to see is that like the sand pile model, it self-organizes to a critical state. By that I mean eventually the bridge model is going to start producing cascades that follow a power law distribution. What we see in the simple bridge model is that if we have a whole bunch of neighboring squares with exactly three players, then by adding one more bridge player we're going to get a sequence of cascades, a huge avalanche. It's this connectedness and interdependence—there are these words again, connectedness and interdependence—that enables the cascade to spread.

I want to take this bridge model and I want to apply it to cost overruns. Suppose something goes wrong on a project; let's think of the Big Dig. Suppose we have a ventilation tube that we realize has to be made a little bit larger. Suppose that some pipes start leaking because this ventilation tube isn't big enough. It could be that this cost increase is isolated; so if this were the case, it would be like the four bridge players just running into open squares. If so, the cost overruns are going to be small. But it could also be that the size of this ventilation tube is such that it requires moving the main drainage pipe. Now we have bridge players moving into new squares and causing people in those squares to disband as well. It might further be the case that moving the main drainage pipe requires tearing up a portion of the road; and so on and so on and so on. There's a song about this; it's called "I Know [an Old] Lady That Swallowed a Fly."

What does this have to do with traffic jams? Again, we're only demanding a somewhat loose metaphorical fit. Higher fidelity models can be constructed that tie up all the loose ends; for the moment, we're just looking for general intuition. Suppose we have a car driving down the road and there are no cars behind it. Imagine the driver of the car hears the song the Eagles's "Peaceful Easy Feeling" on his radio and begins to sing along, but in the process he slows down from 65 miles per hour to 45 miles per hour. If there are no other cars on the road, the fact that the car slows up doesn't have any effect; again, it's like the four bridge players heading to empty squares. But suppose instead the car has six cars directly behind it, and they are a little too closely packed. Now when our car slows up, so must all of the cars behind it. Even worse, even if the Eagles song ends and our driver realizes he's driving down the Santa Monica Freeway at 45 miles per hour, the problem isn't going to go away; because even though he accelerates, it's going to take a lot of time for all six of those cars to accelerate up to speed. It may be sufficiently long that the seventh car back and the eighth car back—which were originally 40 or 50 yards behind the sixth car—may have to hit the brakes, and what's going to happen is now it's going to precipitate further delays and extend the traffic jam.

Finally, let's think about war deaths. Think about the number of squares on the checkerboard with three bridge players as tense regions of the world. If the number of such states is large, then the global political system as a whole can be said to be in a high tension state. If the system is not in a high tension state, adding another player has no effect. If one square gets four players, the cascade likely stops as soon as those people move. The conflagration is contained. The incident goes down in history books as the Pig War, or maybe even the Falklands War, in which approximately 900 people died in a 74 day skirmish between Britain and Argentina.

Other times though, the war spreads, and this was the case with World War I. A potted history of World War I goes as follows: June 28, the Archduke Franz Ferdinand of Austria-Hungary is assassinated in Sarajevo; and it's done by what they think is the Black Hand, a Serbian Nationalist society. Austria-Hungary sends a harsh ultimatum to Serbia. No country would ever accept this ultimatum; and Serbia, as expected—and perhaps desired by Austria-Hungary—rejects the ultimatum. Serbia isn't without allies here; it has strong ties to Russia. Afraid Russia might enter, Austria-Hungary seeks assurances from Germany that they would help defend against a possible Russian-Serbian alliance. After these alliances form, the cascade into war happens pretty quickly. Austria-Hungary declares war on Serbia one month to the day of the assassination. Russia, by treaty, enters to defend Serbia and mobilizes its giant army. Germany responds to Russia's mobilization and declares war on Russia.

Just like the bridge players spreading out, so spread the war. France, bound by treaty to Russia and Austria-Hungary finds itself at war with Germany. By August, Britain and Japan have entered on the side of Serbia, Russia, and France. Why? Britain had no treaty that required it to support France, but it relied on France and Russia to help maintain its naval power. This concern for maintaining naval power led Britain to ask Japan for help against the Germans in waters around China. Japan, eager to show its strength and to establish power over China, obliges, and issues an ultimatum to Germany, that—big surprise—was rejected; so now Japan's in the war against Germany as well. Even countries like Belgium, which intend to be neutral, get dragged into the mix. Why? Belgium finds itself on the road from Germany to Paris, so it gets invaded just because it's in the way. The invasion of Belgium provides yet another reason why the British jump into the war; because it turns out they have 75-year-old treaty with Belgium to protect it.

I'm not saying World War I maps in a one to one fashion to either the sand pile model or the bridge player model, and I wouldn't try to press that. But the chain of events reveals a similar phenomenon: We have a relatively small event that triggers more events because the parts are connected and interdependent. The entire system was poised in a tense state, so what could have been a small event—if not a Pig War, maybe just a Falklands war—became what was called "the war to end all wars," involving upwards of 60 million people and costing the lives of some 20 million people. Once the bridge players started to spread, they kept spreading.

What did we learn? The sand pile model and the bridge player model produce a basic intuition that some systems can self-organize into critical states in which small events trigger large cascades. How does this help us? How it helps us is the following: It enables us to make better sense of the world. We now understand how this combination of connectedness and interdependence implies that the size of earthquakes, traffic jams, cost overruns, and wars may not be normally distributed. If these systems self-organize to critical states, then it's possible that the events will be distributed according to a power law, with lots of small events and these rare cataclysmic events. It also teaches us that as the world becomes more connected and more interdependent, we may be more likely to see large events and so we should probably do something about it.

The fact that we can build models that produce these big events gives us the chance that in some cases we might be able to prevent big events from occurring. We have to be careful to not allow systems to evolve without structure or constraints; because if we don't put in checks to make sure that tensions don't build up, then we might have big events. I want to give three examples. Remember we talked about traffic jams and how they can have long tails. We could stop those long tails if we would just limit access to roads when they became too crowded, because less-crowded roads can't produce these big events. Cities like London limit access to the city by charging an admission fee; this prevents huge tie ups.

Second, let's think back to our original story involving the large event that was precipitated by a single football fan forgetting his camera bag. Events like that can be prevented if connectedness is reduced. How might this be accomplished? Security might develop procedures to seal off portions of the airport very quickly. This prevents the cascade from occurring; it sort of walls off how far grains of sand or bridge players can flow.

Finally, in the case of world wars, it would seem important that we keep a close eye on systems of alliances, as well as on overall levels of tension between countries. One good way to stop a cascade is to alleviate tension. To prevent catastrophic events, we may need to move beyond loose journalistic accounts that "the international scene is tense" to more elaborate understandings that capture the networked alliances, grievances, and so on. We might then work to alleviate tension in critical regions. We have no guarantee that we can predict or prevent war, but it's better to try and understand these processes through modeling than to sit back and wait for the cascades to occur. The same even goes for earthquakes. If we can identify fault lines that are nearing critical states, we might create small earthquakes to alleviate those tensions.

Central to this entire analysis—to all of our thinking here—is the idea that the distribution of outcomes (the large events) depends on the complexity of the system. This idea that unpredictable events are not random but the output of complex processes represents a fundamental shift in how we think, and that's where we're going to turn next.

Complexity versus Uncertainty
Lecture 11

We focus on the difference between thinking of events as random and thinking of them as the output of a complex system and why this distinction matters.

What is the difference between thinking of events as random and thinking of them as the output of a complex system? The New York Stock Exchange is a complex adaptive system, and it produces large events. These large events are mostly crashes, but there have also been days in which the market has made extraordinary gains. We can think of these fluctuations in either of two ways. We can think of them as random events. We can think of them as outcomes of complex systems. In this lecture, we focus on why this distinction matters.

Our analysis consists of three parts. Examining the sources of randomness, including complexity. Examining the difference between randomness that comes from complexity and other sorts of randomness. Examining how the way we think about interventions changes when we take up a complex systems perspective.

Where does randomness come from? There are three standard accounts of the source of randomness. Randomness can be engineered. Randomness can be caused by another randomness. Randomness can be a fundamental property. Complexity theory offers a fourth possible explanation: interdependent rules. Recall our discussion of cellular automaton models, where we considered strings of holiday lights and each light followed a rule. If we begin with just one light, the light's sequences of ons and offs, over time, will appear random, and we would not be able to predict with any accuracy what the next state would be. A simple cellular automaton rule produces randomness, and what we see as fundamental randomness may be the result of simple interacting rules.

This insight—that complexity produces what might appear to be random outcomes—leads us to our second point: how randomness that comes from complexity differs from other sorts of randomness. First the (obvious) intuition. Think back to our lecture on diversity, where we talked about how complex systems adapt. This implies that we have no guarantee that the distribution of outcomes in the future will be identical to the distribution of outcomes today. In the formal language of statistics, this is referred to as nonstationarity. A process in which the distribution of outcomes does not change, on the other hand, is said to have stationarity.

Where does randomness come from? There are three standard accounts of the source of randomness. … Complexity theory offers a fourth possible explanation: interdependent rules.

Why is the distinction between stationarity and nonstationarity so important? Consider the case of Long-Term Capital Management, a hedge fund that failed and failed big due to unexpected market complexity. For another example of how stationary thinking in a complex world leads to tragedy, we need look no further than the 2008 home mortgage crisis. The author and decision theorist Nassim Taleb refers to large, hard-to-predict events such as these as black swans.

This leads us to our third point, which is the idea that complexity not only creates randomness but also has implications for how we think about interventions. Suppose we have developed a potential cure for high blood pressure that might lower blood pressure by 10 percent in approximately 30 percent of patients, compared to a placebo that lowers blood pressure by 10 percent in only 5 percent of patients.

Let's contrast three views of the drug efficacy data. We could adopt a uncertainty mind-set and think of the success of our elixir as a random event. Doctors take a conditional probability approach: They think of the outcome as conditional on the patient. Finally, we can take a complex systems view of the intervention. If we focus on just two implications of complexity thinking, even if the elixir works, we have no guarantee that it will continue to work in the future. On the other hand, if something doesn't work, that doesn't mean it won't work the next time.

We have talked a lot in these lectures about how much complexity there is in the world. Economies, political systems, social networks, ecologies, and even our brains can be thought of as complex. The outcomes of those complex systems fluctuate. It may be easier to think of those fluctuations as random events, but as we have seen in this lecture, if we ignore the complexity that underlies the fluctuations, we can produce large events and we can perform improper interventions. ■

Suggested Reading

Holland, *Adaptation in Natural and Artificial Systems.*

Questions to Consider

1. Do your financial advisors give you advice based on a stationary model of randomness or a nonstationary model? If the former, are you concerned?

2. Even though we cannot predict the future, use ideas from all of the lectures to explain why we might be able to predict the distribution of future events.

Complexity versus Uncertainty
Lecture 11—Transcript

In this lecture, we're going to focus on a conceptual issue: the distinction between thinking of events as random and thinking of them as the output of a complex system. Let me explain what I mean. The New York Stock Exchange is a complex system. It consists of traders who have diverse incentives, strategies, and information who are adapting in response to price changes and information. Like many complex systems, the New York Stock Exchange can produce large events. Let me just describe three.

On October 28 and 29, 1929, the stock market fell a combined 23 percent. This drop presaged the Great Depression, which was an over-a-decade-long worldwide economic slump. Number two: Monday, October 19, 1987, the market fell an astonishing 22 and a half percent for no apparent reason. The financial service industry took a beating, and the New York real estate market went into a price spiral. Number three: During the week of October 6, 2008, the market fell a total of 18 percent. This led to an unprecedented government bailout of the entire financial services industry.

These three large events are all crashes. There have also been days in which the market has made extraordinary gains. On March 15, 1933, the market jumped over 15 percent; so much for bewaring the ides of March. These big jumps often follow close on the heels of the big drops. For example, on October 21, 1987—two days after the 22 percent fall—the market jumped up 10 percent. And twice in October, 2008—after that 18 percent fall—the market jumped more than 10 percent.

We can think of these fluctuations in either of two ways, and this is the key point of this lecture: We can think of them as random events, or we can think of them as outcomes from complex systems. In this lecture, we're going to focus on why this distinction is so important. The analysis we're going to put forward is going to consist of three parts. First what I'm going to do is I'm going to discuss sources of randomness, and complexity is going to be one of them. Second, we're going to examine how randomness that comes from complexity differs in kind from other sorts of randomness. Third, we're going to briefly discuss how a complex systems perspective—how thinking of these things as outcomes of complex systems—changes how we think about interventions. This last part of the lecture is going to be a lead in to the final lecture that considers the broader question of how do we harness complexity once we understand it?

Part one: Where does randomness come from? In the last lecture, we talked about the central limit theorem; and if we add up random events, we get a bell curve, a normal distribution. For the most part, that discussion took the randomness as given; we just sort of assumed it. It left open the question of where randomness comes from. That question is more difficult than you might think. I want to begin with three standard accounts of the source of randomness.

When statisticians teach randomness—and I used to do this—they do so with models that engineer random outcomes, such as when a lottery randomly selects numbered ping pong balls from an urn. You're engineering the randomness by creating a machine—ping pong balls and an urn—that produces it. This sort of randomness—engineered randomness—has been used for a long time; it was used in ancient Athens to select members of the Assembly. When we engineer randomness—when we flip a coin, or roll a die—that's fine; but the randomness that we see in the social world isn't really of this type. In physics and biology, we have very little evidence to suggest that quarks or DNA are flipping any coins; that somehow the randomness was engineered. We have to look elsewhere. Engineered randomness is useful for teaching, but it isn't what we see when we look out there.

What's another source of randomness? The second source of randomness is disturbances. Suppose I want to drive from Chicago to Evanston. It's about 12 miles; it typically takes about 25 minutes. But any given trip may take more or less time depending on how you hit the lights, whether there's a delivery truck blocking a lane on Sheridan Road, etc.; a bunch of reasons. Any one of these disturbances can be seen as a random event. But wait, you might ask, what's the cause of the disturbances? The disturbances come from random events. Your delay is caused by some other person's random decision to take a drive rather than take the bus, or someone's decision to comb his hair at a red light.

The explanation that randomness is the cause of randomness reminds me of an old story, and one version of this story you can find in Steven Hawking's book *A Brief History of Time*. My version's a little bit different. In my version a child approaches a wise man and says, "What holds up the world?"

The wise man rubs his chin and answers, "The world is held up by a very strong man."

The child then asks, "But sir, what holds up the man?"

The wise man replies, "The strong man stands atop a giant turtle."

"But what holds up the turtle?" the boy asks.

The wise man says, "The turtle stands atop another turtle."

"But, what …" starts the boy, but the wise man interrupts.

"Don't bother," he says. "It's turtles all the way down."

The same might be said of this second explanation of the source of randomness: it's randomness all the way down. Think back to the last lecture where we discussed the central limit theorem. The theorem states that if I have bunch of independent random events that get added together, the result's going to be a normal distribution; a bell curve. Recall also that it doesn't matter if the events themselves are normally distributed. All the theorem requires is that the variables have finite variance; this means that their distributions can't be too extreme. Therefore, if we do have turtles all the way down, what we're going to get is a bell curve.

Finally, we can think of randomness as a fundamental property. In quantum mechanics, ultimate values are thought to be probability distributions. The value of that distribution becomes actualized when you measure a particle; so before a particle is measured, it's a distribution, and when you measure it you get a value. That may sound crazy, but a whole lot of quantum physics sounds crazy.

It should come as no surprise that many scientists are dubious about claims of this fundamental randomness. Einstein, in particular, questioned the fundamental randomness in quantum mechanics, and he said—he has a famous quote—"God does not play dice with the universe." That's true; God may not, but something is, and we only have these three explanations at the moment. We have: Randomness can be engineered; it can be caused by randomness; or it can be fundamental. None of these explanations is particularly convincing. Complexity theory offers a fourth possible explanation: interdependent rules.

Now I want to step back. Think of our discussion about cellular automata models. This was the case where we thought about strings of lights, or little triangles of lights, where each light was connected to two other neighboring lights. Each light could be either on or off; and each light followed a rule, and the rule gave the state of the light—either on or off—depending on the states of the neighboring lights. In that lecture—in that previous lecture— we saw how a very simple rule could create a blinker. I want to consider a different, slightly more complicated rule. The following rule's going to tell me whether a light is off in the next period.

A light is going to be off in the next period if either a) it and its neighbor to the left are both on; in a long string of lights, if the light and its neighbor to the left are both on, then it's going to go off. Alternatively, if the light is off and both of its neighbors are in the same state—if they're both off, or if they're both on—then it will go off. If the three lights are in any other configuration, the light is going to be on in the next period. The only way it goes off is if it and its neighbor to the left are both on, or if it's off and its two neighbors are doing the same thing.

Let's imagine that we have a long, long string of lights, and we begin with just one light in the center. What's going to happen is this rule is going to produce an expanding sequence of lights, and they're going to be blinking on and off in all sorts of irregular patterns. If we trace the path of the light that was initially on and analyze its sequences of ons and offs over time, that sequence will be random. What this means is the following: This means that if we look at the sequence of ons and offs, we're not going to be able to to predict with any accuracy at all what the next state is going to be.

This model was developed by Steven Wolfram—I talked about him before—who is a physicist, and he uses this as a model to produce randomness. Let me explain what "randomness" means here in a second. If I flip a coin 10 times, I might get heads-tails-heads-tails-tails-tails-tails- heads-heads-tails. Knowing the first 8 flips isn't going to tell me anything at all about the 9th or 10th flip; there's no pattern. The sequence is completely random. But if instead I were to ask somebody off the street, "Hey, write down a random sequence of heads or tails," I probably won't get a random sequence. I might get something like heads-tails-heads-tails-heads-heads-tails-heads-tails-heads; and here's why: When people write down random sequences—if you ask them to—they tend to underestimate the length of stretches of the same flip. Therefore, if I have a person writing down what's supposed to be a random sequence of heads or tails, if they write down three heads in a row, then I can guess with pretty high probability that that person's going to write down a tail next. If I make that prediction, I'm going to be right more than half the time, and that's because the sequence isn't random; whereas if you're flipping coins if you get three heads in a row, it's equally likely that you're going to get a heads or tails in the next period. What I've told you so far is that this really simple rule—this cellular automata rule; the lights in the long string—produces perfect randomness.

So what? So what? This is incredible. What we're getting is fundamental randomness; so we can actually see this fundamental randomness in the world; that could be the result of simple interacting rules. This explanation that didn't make much sense—to say that maybe there's just fundamental randomness—now we can suddenly get; we can say that there could be simple interacting rules that give us this fundamental randomness. It means that we don't have to have turtles all the way down; it means we could have the bottom turtle standing atop a complex system of interacting rules. This insight—that complexity produces what might appear to be random outcomes—is, like most breakthrough ideas, both intuitive (but it's sort of obvious) and it's profound; and it leads us to our second point: how randomness that comes from complexity differs from other sorts of randomness.

But first, I want to talk about the obviousness of this intuition. Interdependent rules should be able to produce randomness. Remember from a previous lecture we talked about the Game of Life, and we said that the Game of Life—which is a simple set of rules—can produce a universal computer. They can do anything; simple rules can produce almost anything, so randomness shouldn't seem that remarkable. It's sort of intuitive, given what we've learned so far. But why, then, is the idea profound? We have to think back to our early lectures. In those lectures, and in particular in the lecture on diversity, we talked about how complex systems adapt; how levels of connectedness, interdependence, and diversity, and rates of adaptation are in constant flux. This means that we have no guarantee that the future will look like today, because everything's churning. It means in particular that the rules that people follow may change; so we have no guarantee that the distribution of outcomes in the future will be identical to, or even similar to, the distribution of outcomes today.

In the formal language of statistics, this is referred to as nonstationarity. A process in which the distribution of outcomes does not change is said to be stationary. A person's mental state—my mental state—which is going to depend on the connections and interactions of billions of neurons, a very complex process, is going to be nonstationary. The distribution of mental states that I wake up with next year won't be the same as the one I woke up with this morning.

Why is this distinction so important? Why is it so important to make a distinction between stationarity and nonstationarity? Here's why: If we assume stationarity in a nonstationary world, we may be in for a shock. Let me consider a specific case, and this involves Long Term Capital Management. Long Term Capital Management—more popularly known as LTCM; that's the acronym—was a hedge fund founded in 1994, and on its board were two brilliant economists, Myron Scholes and Robert Merton. Scholes and Merton won the Nobel Prize in 1997 for research on how to price assets. Long Term Capital Management traded in derivatives. What this means is that they made bets about what would happen to the prices of different commodities and currencies. At the core of their trading strategy was the idea that markets equilibrate in predictable ways. When we say predictable, that means a certain level of stationarity. They assumed that certain long-term ratios would persist.

Let me abstract away from all this fancy finance talk. Imagine a country in which people love to hike; and if you hike, you're going to wear out socks and you're eventually going to wear out shoes. Socks wear out faster than shoes, but we'd expect that sales of these two commodities to move in a fixed ratio; so supposed you'd buy six pairs of socks for every pair of shoes. To keep this story simple, let's suppose that shoes cost about six times as much as socks. That's going to mean in the long run, the amount of money that's spent on socks is going to equal the amount of money spent on shoes. You think of the price of a stock as just being a multiple of the expected sales, so that's going to mean that the price of a stock in the shoe industry is going to be the same as the price of a stock in the sock industry. It depends on what people will pay for the stock, when you think of an actual stock price, and that takes into account all sorts of other influences like people's beliefs about market growth, brand loyalty, pace of innovation, and so on.

Because stock prices—and sock prices in this example—are going to depend on what other people think are going to happen to those prices, that means that in the short run we can have little positive feedbacks; that means prices can be incorrect. Suppose that the value of the stock of the sock industry in this case starts to rise relative to the value of the shoe industry because people think it's going to rise; this is what I meant by a positive feedback. People are going to invest in socks mainly because other people are investing in socks, and the price is going up. As a result, the value of the socks is going to keep going up because of the fact that people are investing in them and it's going to look like an even better and better and better investment.

But this sequence of events has little to do with what we call market fundamentals. Market fundamentals say that in the long run income in the two industries—the sock industry and the shoe industry—should be approximately the same. Because these two things are used in tandem, this is going to mean that this bubble—this sock bubble—is going to have to crash. This is, in effect, what Long Term Capital Management bet on: They bet that prices of currencies—not socks, but currencies—would get back in long run equilibrium. The problem is the bets that Long Term Capital Management made didn't have an infinite shelf life. They only applied during a fixed window; so they could only call these bets during a certain period of time, and when that window expired, Long Term Capital Management was going to lose. In this case, they lost big.

The experts at Long Term Capital Management knew about bubbles. They also knew that historically bubbles only lasted so long, which is why they were so confident that their bet was going to pay off. However, what happened was financial markets had become more interdependent; so that means the distribution wasn't stationary. The financial markets were in a state of flux with increasing connectedness and interdependencies. Specifically what happened was this: When they were making this bet, the Russian ruble collapsed. This resulted in increased risk aversion on the part of lenders. Owing to the fact that markets were more connected than they'd been in the past, investors temporarily demanded a little bit more liquidity; so it was almost if for the winter people wanted eight pairs of socks per pair of shoes rather than the usual six. A huge bet—with borrowed money I might add—that the ratio would return to six immediately, in the middle of February, suddenly became a bad bet. In the end, the market fundamentals did return to where they should have been—it returned to where the socks and the shoes had the same price—but it was too late. Long Term Capital Management didn't have the money to pay off its debts—the money they borrowed to make these giant bets—and the government bailed them out, fearing that if they didn't we might see a more general financial collapse.

Let me give another example of how stationary thinking in a complex world leads to tragedy. Here we need look no further than the 2008 home mortgage crises. First some background here, I think, is very important. A hundred years ago if you wanted to buy a house, you needed 50 percent down—that's right, 50 percent—and when you got that loan, the paper (the title) was held by your local bank, which means that they took all the risk; that's, by the way, why they demanded 50 percent down. Now when you get a loan, the originator of the loan sells that loan off to someone else who bundles your loan with that of many others. Bundling is great because it reduces risk; and the logic here is the logic of the central limit theorem: Some of these loans are going to get paid off, some are going to fail, but if you bundle enough loans together—some good ones, some bad ones—what you're going to get is, on average, very little risk. You know what you're going to get; you're going to get the middle of that bell curve.

What you have is banks and other financial institutions like Fannie Mae and Freddie Mac that had bundled these loans, and then they had to figure out what these bundles were worth; they had to price them. They are two methods that are used to bundle loans: The first one is called mark to market, and the second is called mark to model. In the first approach, what you do is you go out and you sell a portion of the bundle; so you take a representative sample of the bundle and you see what it fetches. This allows you to sort of mark the value of the full bundle to what the market will pay. Alternatively, you can mark your bundle to model. This means what you do is you rely on some sort of mathematical or statistical model of how the world works to evaluate what the bundle's worth. This is risky; it's incredibly risky if you assume stationarity.

One reason why we had this giant collapse in 2008 is because the models that these investment firms used to mark to model were wrong. Why were they wrong? In the past, foreclosures tended to be regional. Remember the second crash I began this lecture with, the 1987 stock market crash. In that crash, the New York real estate market went into a price spiral. But the economy wasn't as connected then as it is now, and home values weren't as interdependent, so it was contained. The models that they were using in 2008 underestimated the connectedness; they underestimated the complexity of the housing market and they thought that if there was a collapse it would be contained, which meant that housing values would be independent so that when we added them up, we'd get a nice bell curve distribution.

Second problem: The assets that were being bundled had changed. The models also didn't take this into account: A combination of government policy and market incentives had resulted in an enormous number of what are called subprime mortgages. Briefly, a subprime mortgage requires very little money down—nowhere near 50 percent—and often it requires no verification of income or assets. Why would somebody make a loan like that? The reason why: they pay higher returns, and that higher return is basically made up for by bundling; and so the risk goes away by bundling these things. By 2007, 30 percent of new mortgages could be put in this subprime category. It seems crazy to make these loans, and they seem very risky; but when housing prices are going up, you have no problem at all, because even if you lose your job, as long as the housing prices are going up you can sell your house at a profit in the rising market and just buy another one.

This wasn't necessarily a bad idea. Allowing people with very little wealth to buy houses allowed these people to share in the rising housing market; so this was something the Federal government had mandated. We can think of these subprime mortgages as an adaptation within the complex system. They were an adaptation that was intended to benefit both the people who took out the loans and the financial services industry that wrote them and bundled them; because the bundlers were getting a higher return at a cost of what they thought was slightly more risk because they were using central limit theorem sort of logic.

These subprime mortgages became a cancer when the market went south. Just a slight hiccup in the economy and a no money down mortgage will go bad. For a 50 percent down payment to go sour, the economy just has to completely tank. Now, again, you might think: But these financial services companies, they hire the best and the brightest; how could they be so stupid as to not think that many of these subprime mortgages could go bad? The point is they weren't stupid; they were smart. They bought insurance, and the insurance came in the form of what are called credit default swaps. These are investments that pay off if the bundles turn bad. Someone had to sell them these credit default swaps; this insurance. Who did it? It was giant insurance companies; they stepped up to the plate. Or, in other times, the firms self-insured; and that's an even riskier thing to do. They put a little more money away just in case things went bad.

The problem is they didn't put enough money away. Once people started to default, housing prices fell. Owing to all the no money down mortgages, this meant that a whole lot of people's mortgages exceeded the value of their houses. What we get is a positive feedback: As people default, they cause housing prices to fall, creating more defaults, which creates a fall in housing prices, which creates more defaults. Second, as banks and other financial institutions realized they were holding bad paper—bad loans—they decided they needed to hold more cash to balance their portfolios, to keep themselves financially solvent. This tightening of credit led to—you guessed it—less money for people to buy houses with, which leads to lower housing prices, which leads to more defaults, which leads to investment houses needing to hold more cash, which leads to higher loan rates, which leads to more defaults; another positive feedback, a second positive feedback.

Notice the role that complexity plays in this version of events. Financial markets were more complex than analysts anticipated owing to greater connectedness and interdependencies, as well as market adaptations in the form of these subprime mortgages. This led to a mispricing of assets; and from here, positive feedbacks take over, and we get this disastrous large event. It's terrible.

Wait, wait, wait; you might say, "Wait a minute!" Earlier on in these lectures, I gave a model of emergent firewalls. Remember that model? You had some banks that learned to be safe while others took risks, and the result was these emergent firewalls and this robust financial system. Was that model wrong? And if so, does that mean that we just can't trust these simple complex systems models to help us find our way? No, that model wasn't wrong; in fact, that's the reason for this lecture. That model also assumed stationarity. It assumed that the failure rates stayed constant so that the banks could learn what to do; and learning what to do meant building firewalls. It also assumed that the connections between the banks stayed fixed. But both of those assumptions failed to hold in the real world; the real world was not stationary, and that lack of stationarity was what caused the problem. The author and decision theorist Nassim Taleb refers to these large, hard to predict events, such as the 2008 collapse or the Long Term Capital Management bailout, as black swans. Failing to recognize complexity as the source of randomness is a leading cause of black swans.

This leads to the third point: the idea that complexity not only creates randomness but somehow has implications for how we think about interventions. I want to leave the financial sector for a second and consider disease and disease prevention. I want to do a little thought experiment; I know I do a lot of these, but they're really helpful. Let's suppose that we've developed a potential cure for high blood pressure, and it involves some mixture of ginger, ginkgo, and vitamin B12; something like that. How are we going to test this magic elixir? We're going to hire an independent research laboratory to conduct clinical tests. The results of this might look as follows:

We might find that our elixir lowers blood pressure by 10 percent in 30 percent of patients, compared to a placebo that may lower blood pressure by 10 percent in only 5 percent of patients. These tests would also look for side effects: the bloating, the nausea, the bushier eyebrows; that sort of thing. By the way, this last number about placebos is realistic. Placebos almost always prove effective because there's what we think of as regression to the mean. There's a tendency for complex systems to sort of recalibrate. If someone has high blood pressure, because the body's a complex system, if we test them in a few weeks, it's likely that their blood pressure will move back towards normal; it's less likely, then, it's going to continue to spike.

I'm going to contrast three views of this drug efficacy data. First, we could adopt what I call a pure uncertainty mind set; and we can think of the success of our elixir as a random event. If that were the case, we'd just keep giving the patient the drug saying, "Look, it works 30 percent of the time, eventually it's going to work." It would be like rolling a die saying, "Eventually we're going to get a two and you're going to be cured." But that's not how most doctors think. They take what we call a conditional probability approach. They're going to say the outcome for a particular patient is conditional on the patient. Thus, on a given patient, the drug either works or it doesn't work. So if the drug didn't work once, they wouldn't try it again; they'd chalk up the difference to genetics, to some weird body chemistry, to some conditional event, or some conditional property of the patient. But third, we could actually take a complex systems view of the intervention. If we do this, we think of the intervention in terms of genetics, body chemistry, and lifestyle. We think of the person—their body—as a complex system. Now the success of the intervention, if it were, is not a random variable but the outcome of a complex process. That can be many different things; so let's just think about two implications of complexity thinking here, and both are fairly provocative.

The first one is this: Even if the elixir works, we have no guarantee that it's going to continue to work in the future. For example, you can go on a diet and initially lose 10 pounds; and then you may find that your body weight has crept back up even though for the most part you've stayed true to the diet. Why is that? The reason that happens is physiologists have found there's a lot of evidence that you have a metabolic set point. In other words, your body as a complex system maintains a constant metabolic rate, just like the beehive maintains at a constant temperature. The more muscle mass you have, the higher that rate is going to be, for example; but changing your metabolic set rate isn't very easy. The body has a number of built in homeostatic feedbacks; again, those bees buzzing away in the hive. When we diet, our cells basically send signals to the brain that say, "Get me some pizza and ice cream and do it now" (they don't say that exactly, but in my case, it's actually pretty close). Our brain then does two things: it slows our metabolism to conserve energy, and it sends out a hunger message making us just ravenous. That's why when we diet, we feel sluggish; and then we go on binges and then we find out that we're back to our original weight.

Unfortunately, it appears a lot easier to lower our set point than to raise it; because once we build fat cells they can continue to send out those messages and we can never get rid of them, we can only shrink them. So a complex system approach implies that even if something works, we should be prepared for it to work only for a short time, because the system's going to react to it. It also implies that if something doesn't work that doesn't mean it won't work next time; it's literally like rolling the die again. This may be especially true if you're curing psychological problems and changing behavioral patterns. In fact, the American Lung Association claims that every time you try to quit smoking you increase your chances of quitting in the future. This suggests that each time you try you change the connections and interdependencies in your brain patterns and in your daily behavioral patterns. What you have is a fight between these negative feedbacks that want your brain and behavioral patterns to return to the status quo—to smoking—and these positive feedbacks that are trying to tip you to a new smoke-free life. It's almost like our very first example with the ball and the basin: the more you push that ball, the more likely you are to get it out of the basin.

We've talked a lot in these lectures about how much complexity there is in the world: Economies, political systems, social networks, ecologies, and even our brains can be thought of as complex systems. The outcomes of those complex systems fluctuate. It's easier—it's comfortable—to think of these fluctuations as just random events; but as we've seen in this lecture, if we ignore the complexity that underlies the fluctuations, then we're not going to understand where these large events come from, and we're not going to be able to perform the proper interventions.

By definition, complex systems are in constant flux. They adapt; they change; they cannot be told what to do. At best—the best we can hope for—is to harness them and steer them. Our capacity to do even that is going to be the subject of our final lecture.

Harnessing Complexity
Lecture 12

If a system is complex, can we intervene productively? Can we reduce complexity? Would we want to?

W hy would someone want to learn about complex systems? I came up with two reasons, which are also the reasons why I continue my own research into complex systems. Complex systems are inherently interesting, producers of amazing novelty, and not clean and simple, which makes them a lively playground for the mind. Complex systems are where the action is. The fundamental challenges of our time are all complex.

In the first 11 lectures, we focused primarily on learning the basics of complex systems. In this last lecture, we turn to the takeaways. What have we learned from this brief foray into the study of complex systems? How can it help us choose better courses of action, or even to make sense of the complex world around us? I think of this last lecture exploring the space between "lion taming" and "poking the tiger with a stick." We cannot hope to control complex systems through interventions. At best, we might learn to harness and respect complexity.

We will begin by describing a noncomplex system of making choices that is known as decision theory. This will provide a benchmark against which to compare the lessons we learn from complex systems. The canonical decision theory model of how to make choices can be described in a few steps. The first step is to determine a set of options. The second step is to determine the payoff of each option in each state of the world. The third step is to compute the probability of each state of the world. Once you have written down all of the options, all of the possible states, and the payoffs of every option in every state, then you can make a rational choice.

This canonical decision-making model works great if, for example, you want to decide which computer to buy, but it is not a very useful model for determining what to do in a complex system. Let me give four big reasons. The standard decision-making model does not take into account the behavior of other interested actors. The standard decision-making model translates complexity into uncertainty. The standard decision-making model is all exploitation. The standard decision-making model focuses on a single outcome, not on system properties. Therefore, the model takes no account of what the system might be like as a result of your action.

The first step toward effective action in a complex world is recognition. Not everything is complex. Some systems are linear and predictable, though these are not the same thing. An effect is linear if we get a straight line when we plot it. An effect is predictable if we know what will happen. An effect can be linear and predictable or linear and unpredictable. An effect can also be nonlinear but predictable. Finally, we can have nonlinear effects that are unpredictable. Complex systems often produce these sorts of effects.

Systems that are not complex can be controlled. We can figure out what to do best. Situations that are complex require an awareness of the parts that make them complex so that you can keep an eye on key attributes. Once complexity has been recognized, we at least have the hope of harnessing it, of taming the lion.

Let's think about how we might harness complexity to do good. Our first step will be to think in terms of

Ralph Waldo Emerson wrote, "As soon as there is life, there is danger."

the attributes of choice variables or levers, which we can choose to increase diversity or decrease interdependencies. Let's start with diversity, where our first insight is to encourage diversity—but not too much. Without some source of diversity, selection will drive systems toward pure exploitation, which can be dangerous. Diversity also prevents error. Another bit of advice is to keep an eye on the tails. Next, let's think about selection mechanisms as a lever, where our second insight is to be careful how you define goals and incentives. Then let's consider interdependencies, where our third insight is to not become so obsessed with making small efficiency improvements that you push a system toward a critical state. Last but not least, let's look at connections, where our fourth insight is to search for potentially synergistic links and cut those that limit responsiveness.

These lessons are easier said than done, but notice the coherence among them. Synergistic links exploit diversity and positive interdependencies. Selection tempers diversity to help balance exploration and exploitation. Pulling back on efficiency a bit to allow some slack not only ensures robustness, it promotes greater innovation by allowing diversity.

Let's do an empirical test to see what types of organizations prove robust. Let's look at organizations that have been around for more than 500 years; this group includes more than 40 universities and around 125 or so businesses. They show that a little slack is a good thing, if you want robustness. In Jenna Bednar's book *The Robust Federation*, she makes the point that robust federated governments also need space for a little slippage.

Complex systems are inherently interesting, producers of amazing novelty, and not clean and simple, which makes them a lively playground for the mind.

Many of the core insights we have discussed in this lecture seem straightforward, but let's just recognize how they depart from the standard idea of command-and-control optimization. If you take a command-and-control optimization approach, you set incentives with only outcomes in mind. Command-and-control optimization is not wrong, but it is a poor approach if we are hoping to thrive in a complex world. Our goal should be to push our understanding of these systems. Only through understanding can we move from a position of poking the tiger with a stick to one where we are taming the lion. ■

Suggested Reading

Axelrod and Cohen, *Harnessing Complexity*.

Questions to Consider

1. Suppose you intervene in a complex system by adjusting the diversity level. How might you know that, rather than taming the lion, you have just poked the tiger with a stick?

2. Given what you have learned about complex systems, do you believe that the success of any one person depends more on their individual attributes or on context—on where that person happens to be situated?

Harnessing Complexity
Lecture 12—Transcript

Welcome to our final lecture. Before I wrote the first word of the first lecture in this series, I asked myself: What would someone most want to learn about complex systems? Why would they want to listen to a course about complexity? I came up with two reasons (I might add that these are the two reasons why I continue my own research into complex systems): The first, quite simply, is that complex systems are inherently interesting. Whether it's emergent phenomena like consciousness and culture, or it's the fact that these systems are both robust and prone to producing these large events, you don't even have to be in the mood to be amazed to find complex systems amazing. They can also produce amazing novelty; and they can turn on a dime, so to speak, by producing these phase transitions. In the end, complex systems are fascinating because they're not clean and simple; and this makes them a lively playground for the mind.

Second, complex systems are where the action is. The fundamental challenges of our time—managing climate change, maintaining ecosystems, regulating financial markets, preventing the creation of conditions that spawn terrorists or produce poverty—all of these are complex. Not in some loose metaphorical sense but in the formal sense we've defined in this course: they involve diverse interacting adaptive entities whose micro-level behaviors produce macro-level patterns to which they in turn adapt, creating new patterns.

In the first 11 lectures in this course, we've focused primarily on the basics of complex systems; we did basic blocking and tackling, so to speak. We named the parts and attributes, and reveled in all of the ideas: self-organized criticality, universal computation, tipping points, homeostasis, power laws, and small worlds. In this last lecture, we're going to turn to some takeaway points. What have we learned from this brief foray into the study of complex systems? How can it help us choose better courses of action, or even to make better sense of the complex world around us? I think of this last lecture as exploring the space between lion taming and poking the tiger with a stick. We cannot hope to control complex systems through interventions. At best, to borrow a term introduced by Bob Axelrod and Michael Cohen, we might learn to harness complexity; we might learn to tame the lion. At the same time, we need to respect complexity. An actor in a complex system controls almost nothing but influences almost everything. Let me repeat that: An actor in a complex system controls almost nothing but influences almost everything. Attempts to intervene may be akin to poking the tiger with a stick.

I want to begin by describing a non-complex system way of making choices that's known as decision theory, and this is going to give us some grounding. Decision theory predominates in business, government, and the nonprofit sector. We're going to see why this approach doesn't work in complex environments. It's going to provide a benchmark in which we can compare the lessons we've learned in this course about complex systems.

The canonical decision theory model of how to make choices—this is the one that's taught in business schools, medical schools, and public policy programs—can be described as follows: You have this decision maker who has this set of options before her. Time has two points: there's now, and then there's the future. In the future, there's some state of the world that's going to be revealed. For example, if you're contemplating buying a car, the possible states of the world might include moving to New York City, where you don't need a car; or it might include having more children, which would mean you'd need a different type of car, one with more seats. The second step in making a decision in this way is to determine the payoff of each option in each state of the world. For example, if the option is to pack an umbrella, and the state of the world is rain, then the payoff of the umbrella will be very high in that state. For each of these actions, and each of these states, you have to figure out the payoff. The third step is to then compute the probability of each state of the world. The final step, then, gives us a cost benefit analysis; we just want to choose the option that has the best expected payoff. This may not be the one that has the highest expected monetary payoff, because we might care about risk.

Once we've written down all the options, all the possible states, and the payoff of every option in every state, we sit back and we make what is thought of as a rational choice; our best possible choice. I want to come clean right now and tell you that I love this model; it's a great model. I teach it to my undergraduates every year. It's extremely useful for teaching students of any age how to frame choices and how to compute the value of each choice and of having more information. For example, you can use this model to determine how much you would pay to reduce uncertainty over future state of the world. Businesses use this model daily; a big pharmaceutical company, for example, will use it to determine which medicines that it has in its pipeline should stay, and which ones they should—in their terms—explode.

I love this model, it's true; but I'm also very aware of its limitations. I love my bike, it's great; but I'm not going to use it to cross the Atlantic. Comparatively speaking, that's what a lot of folks are asking of this canonical model when they apply it to complex systems. This canonical decision-making model works great if you want to decide, "Should I buy a computer?" or "Should I prepay for a lower priced ticket for my next vacation?" But it's not a very good model for determining what to do within a complex system. Let me give four what I think are very big reasons.

Reason number one: The standard decision-making model doesn't take into account the behavior of other interested actors. This shortcoming has long been known, and so the framework has been extended to create something called game theory. Game theory is basically decision theory amended so you can take into account actions of other agents. Reason number two: The standard decision-making model translates complexity into uncertainty. All of the complexity gets folded into some fixed probability distribution over the states of the world. Remember in the last lecture, we saw how assuming this stationarity—assuming there's just some sort of distribution out there—creates problems. We saw that in the collapse of Long Term Capital Management. Reason number three: The standard decision-making model is all exploitation. Remember, this means making decisions or choices based on what you already know, so it doesn't allow for any exploration; it tells you make the choice that gives you the best payoff. But in complex systems, you want to balance exploitation with exploration. Remember, the landscape dances; and if the landscape is dancing, you have to continue to explore. Reason number four: The standard decision-making model focuses on a single outcome, not on system properties such as connectedness, interdependence, diversity, rates of learning and selection, and so on. Therefore, the model doesn't take into account at all what the system might look like as the result of your action. Let me give an example: Your action may reduce diversity to dangerous levels; even though you get a good outcome, you've changed the system in such a way that it doesn't have sufficient levels of diversity.

The first step toward effective action in a complex world is recognition; we have to recognize when a situation is complex and when it's not. Not everything is complex; some systems are linear and predictable. I should add these aren't the same thing; let me differentiate them. An effect is linear; if when we plot it we get a straight line; so the amount of money raised by a sales tax is linear in sales. In contrast, an effect is predictable if we know what's going to happen; so if I take a bowling ball and drop it on my foot, I can predict that it's going to fall and it's going to hurt.

An effect can be both linear and predictable, like the increase in the weight of a wheelbarrow as I add sand to it. An effect can also be linear and unpredictable; what would that be? Orley Ashenfelter, who's an economist at Princeton, has found—and this is sort of cool—that the quality of Bordeaux wines increases linearly with the average temperature in the Bordeaux region of France in September. No really; that's true. But here's the thing: We cannot predict whether September is going to be warm or cool with much accuracy at all; so therefore what we have is a linear effect, but that effect is not predictable. It's also true that an effect can be nonlinear and predictable; so the weight of a rock increases with the radius cubed. If I double the radius of a rock, I increase the weight eightfold; so that's not linear.

Last of all—and here we're getting to where sort of the meat of this lecture is—we can have nonlinear effects that are unpredictable, and this is the domain of complex systems. Complex systems often produce these sorts of effects; remember we saw that when the lake became eutrophic: everything seems fine, and then all of a sudden an unpredictable moment—boom—we have this algae-ridden lake. Systems that are not complex—the shoveling of coal—can be figured out; those can be controlled. We can figure out what to do; we can make the best possible decisions. Situations that are complex—running a middle school, for example—require an awareness of the parts that make it complex so we can keep an eye on the key attributes. Just like the fire ranger must create small fires to reduce big ones, so might a school administrator create outlets to reduce small tensions lest they explode into bigger ones.

Once you're recognized complexity, then you at least have the hope of harnessing it; of taming the lion. However, in the face of complexity even the most noble and well-intentioned efforts may fail. I want to consider for a moment humanitarian efforts. Conor Foley, who's a long-time British humanitarian, wrote a book called *The Thin Blue Line*. In this book he demonstrates that interventions "virtually never"—those are his words—resolve crises. He compares them to performing heart surgery using plaster. It's an apt analogy. Countries, like hearts, are complex adaptive systems. You can't go in with the equivalent of scotch tape and hope to fix a heart that's dysfunctional; nor can you do the same with a country.

With that sobering introduction, let's think about how we might harness complexity to do some good out there. Our first step in that process is going to be to think in terms of the attributes that we've had and think of them as choice variables or levers. By choice variables or levers, what I mean is: think of these as variables under our control. Remember back to our lecture on dials. I want to think of our ability to turn those dials. We can choose to increase diversity or decrease interdependencies.

I want to start with diversity. We've talked a lot about why diversity is beneficial—it's the engine of innovation, contributes to robustness, and so on—so organizations and societies writ large probably should encourage diversity; diversity is key. Yet we also saw that if we let the agents become too diverse, then we're not going to exploit what has been learned; so we have to balance exploration against exploitation. That's our first insight: We want to encourage diversity, but we don't want to encourage it too much. This is a key point; because without some source of diversity, selection is going to drive systems toward pure exploitation, and that's going to be dangerous. Remember in an earlier lecture we discussed how selection drives down diversity; that means that we need some force out there to maintain diversity.

When I say "lack of diversity," this can exist not only in how people look; what's really important here is differences in how people think. It's this lack of cognitive diversity that can lead to collapse. This is often known as "group think." Others refer to this as a "takeover by a dominant logic"; that's the management phrase. For example, in the 1980s, IBM had a dominant logic, and that dominant logic was that mainframe computers were the way of the future. The result was that IBM had this blind spot, and it prevented IBM from seeing—let alone anticipating in any way—the personal computer revolution.

Stories like this beg the question: IBM had really smart employees, so how did this dominant logic come to be? To answer this question, we need look no further than one of our earlier lectures on positive feedbacks. Once we have a dominant logic gain a foothold, a number of positive feedbacks kick in, not the least of which is selection. So people who espouse the dominant logic are more likely to be promoted. This in turn creates an upper management that believes the dominant logic, and that creates even more incentives for the underlings to promote the dominant logic as well. Even without the incentives induced by the organizational structure, people might well fall into a dominant logic. As Eric Hoffer wrote in *The Passionate Mind and Other Aphorisms*, "When people are free to do as they please, they usually imitate each other."

Once a dominant logic takes hold, it becomes the way of interpreting events. Every piece of information gets shoehorned into the dominant logic, making that logic seem even more compelling. The resulting group think can lead to disastrous or near-disastrous outcomes, such as when John F. Kennedy and his staff—remember they were called "the best and the brightest"—brought us to the brink of world war during the Bay of Pigs. Again, I want to reiterate: This does not mean that diversity is always better. Diversity isn't necessarily needed if a situation is well understood. But once a situation is complex, and if the errors could result in catastrophe, then it makes sense to inject diversity at regular intervals, lest your dominant logic has you producing mainframes and bundling up subprime loans.

Diversity also prevents error; and this idea finds its clearest representation in something called Linus's law, which goes as follows: "Given enough eyeballs, all bugs are shallow." What does this mean? This law—which is named for Linus Torvald, the driver of Linux—was coined by Eric Raymond in his book-length essay, "The Cathedral and the Bazaar." What he means by "all bugs are shallow" is that any coding bug can be found if enough people get to look at it. What Eric Raymond does in this essay "The Cathedral and the Bazaar" is he basically compares two different systems for writing code: the cathedral model and the bazaar model. The cathedral model consists of the standard business model where software is built by some restricted set of programmers who work like mad debugging every version, and then they only release the versions when they're perfect, or at least as perfect as they can be made. The bazaar model is more familiarly known as "open source." In the open source or bazaar model, the code is written in full view of the public so anyone—any programmer—can go online; any set of eyeballs can notice a bug and contribute to fixing it. They can do this via blogs, commentaries, or open websites. If you have enough eyeballs, any bug is shallow; any bug can be seen.

Interestingly—and here again we're going to see this fan out nature of complex systems—immune systems work the same way. The more diverse your immune system, the more bugs it can attack. Once your immune system is successful, what does it do? It uses a positive feedback mechanism to produce more of the antibodies to wipe out the virus. Note this logic works in reverse; so the more diverse the virus, the less likely your immune system will get it. Think about HIV: What does HIV do? It mutates and recombines as it spreads; that means it's a diverse set of attackers. Hence, if the immune system stops one version, that doesn't mean it's going to stop the others, because HIV keeps adapting.

Good leaders know the value of diversity and they promote it in myriad ways, because they need diversity to prevent errors, and they need diversity to prevent diverse attacks, such as HIV. So what do they do? They rotate people's jobs and offices, they create parallel work teams, they bring in outsiders, and they hire people with diverse training. Remember we were just talking about the Bay of Pigs fiasco? Following that, John F. Kennedy basically saw the need for more eyeballs, so he started inviting outside experts to sit in meetings and question his own advisors so as to ensure some diversity of thought. Kennedy learned the hard way: without diversity, collapse can be in the offing.

One last point about diversity: Recall our previous discussion about the tail wagging the dog; how the average sometimes matters less than what people at the extremes think. This leads to another bit of advice: keep an eye on the tails. Even if on average people are happy, that doesn't mean that a riot won't occur; all it takes is a few angry people and a lot of positive feedbacks.

Now let's think about selection mechanisms as a lever; we just talked about diversity as a lever, let's think about selection mechanisms as a lever (something we can turn). Evolution selects through reproduction. If you're not fit enough to reproduce, your kind doesn't get to continue to live in the future; you don't continue to exist. Organizations and markets do something similar. When an organization decides to promote, they're creating a selection mechanism. Suppose that an organization says, "Let's select the best individual performers for promotion." This seems like a great idea, but what if the best individuals aren't good team players? The result may be an organization that has a bunch of self-interested individuals at the top, and is therefore destined to fail. Alternatively, an organization can go the other direction, and they can promote people who are just really well liked. This could create another sort of problem: This could create a firm full of really nice people who never challenge anyone.

Determining performance measures is really a tricky business; it's a very subtle business. Let me give an example; this is one of my favorite examples in all of complex systems. Karl Sims, who's a computer programmer, once wrote a computer program that was intended to produce objects capable of locomotion. What he did (this is just great): He encoded the laws of Newtonian physics, and he set loose some evolutionary forces of selection, reproduction, mutation, and recombination; and what it was supposed to do was sort of evolve these creatures that moved. His performance measure was just the distance traveled by the center of mass. He set this program loose and let it run, and to his surprise his evolutionary or computer program—this agent-based model—evolved these huge towers; they just fell over. This seems really odd; it doesn't make any sense. But then you realize, wait: What was his performance measure? What was his selection criterion? What lever was he using? It was move the center of mass as far as possible. Think about it for a second. What did his program do? It came up with a brilliant answer; a brilliant design. It got really tall and fell over. When you do that, your center of mass moves really far really quickly.

How did he overcome this problem? He changed his performance measure; he made it the minimum distance traveled by any point on the object. In that case, the thing that fell over, the bottom of it never moved and it wasn't very fit. Once he changed his selection criterion, his program evolved little things that were swimmers and crawlers and paddlers that did exactly what he wanted. So that's insight number two: You have to be careful how you define goals and incentives in a complex system.

On this same point, there's a saying known as Orgel's Second Law, named for Leslie Orgel who was a British chemist, who's actually more well-known really for working with Francis Crick and he later took up residence at the Salk Institute. Orgel's Second Law is this: Evolution is smarter than you are. The way I'd like to interpret this rule is as follows: Evolution is unsparing in what it tries. It is, to quote Dawkins, "a blind watchmaker." It'll find a solution to almost any problem, but it might not be the solution you anticipated. But smarter than we are? Come on, we listen to the great professors on tape; that can't be true. I'm actually sorry to say, but for all of our human powers of cognition, we're often blinded by these dominant logics. These logics may be based on science, religion, or culture. Regardless, because we have them we don't try everything; we don't waste our time trying to do what we think is impossible. Evolution mindlessly tries every possible combination; so with enough time, it's eventually going to win. Despite all its constraints, evolution often can be smarter than we are, then, right? All we have to do is remember the pony fish's glow; remember how their bellies lit up?

Next lever: interdependencies. Here I want to go back to the book by Axelrod and Cohen that is called *Harnessing Complexity*. They talk a lot about the concept of self-organized criticality, leading up to a third lesson. That third lesson is this: Don't become so obsessed with making small efficiency improvements that you push a system toward a critical state. How can this happen? Let's think: Following the terrorist attacks of 9/11, border security between the US and Canada increased. This slowed traffic between the two countries, particularly the Ambassador Bridge between Detroit and Windsor, Canada. There were some companies—actually auto companies—that relied on just-in-time inventory techniques, and they found that they didn't have sufficient parts to continue operations. Some of those companies had suppliers that were in other countries—in Canada—so this resulted in cascading failures. If you have a system like a supply chain with lots of interconnections, you have to leave some slack; otherwise a simple failure, like a backup at a bridge, can lead to a cascade. Now remember, cascades are the idea that one failure begets another, which begets another, which begets another, which is what we saw in our sandpile model of self-organized criticality.

Notice here—this is provocative—this runs directly counter to the advice we get from optimization thinking; from our decision theory model. It's basically saying, "Don't optimize." It's not quite saying that; it's saying, "Don't optimize fully." It's okay to be 98 percent effective, just don't shoot for 100 percent. If the system is completely independent—if you're making the 21-pound shovel—you're fine. But if the system is interdependent, you want to build in some slack.

Last but not least, let's look at connections; let's think of connections as a lever. When you have a network, what happens is that it's formed by agent-level incentives; so the resulting network may or may not have the properties that we want. The Internet—remember we talked about this—is robust to random attack, even though no one set out to make it so. When we think about a complex system, it's important to ask: What were the incentives, and what connections exist? Do we have the right connections? In comparing the connections that exist and the connections that might be, we can think about possible interventions; so there are only two. One is we can make a new connection; the other is we could sever an existing connection. How can we decide to do each? What might we have learned in this course that tells us how to cut and how to connect?

New connections should be created if they produce synergies. For example, we have physicists who study fractals, and we have biologists who identify fractal structures in the nervous and circulatory systems. A physicist who connects to a biologist might produce a synergism; and in fact, path breaking work by Geoffrey West—who's a physicist at the Santa Fe Institute—did exactly that. He and his coauthors—a mix of biologists and physicists—

were able to derive theorems that explain these sort of fractal scaling phenomenon in species. For example, they can explain how metabolic rates in animals scale with size; so it's this great synergy. Ron Burt, a sociologist at the University of Chicago, refers to these sorts of interdisciplinary actions as filling in structural holes. Robert Putnam of Harvard refers to these as bridging links. Regardless of the terminology we use, what we see is the value of creating links between knowledge domains that might not have existed from independent incentives.

On the other hand, sometimes there are links that exist that we should cut, because these links discourage innovation. This idea of severing links underpins what some people call a greenfield strategy. Perhaps the most famous example of a greenfield strategy is the Saturn Auto Company, which was funded by General Motors but which operated completely independently, at least initially; there were very few connections to the main organization. When I say greenfield, what I mean is you create a separate company that's disconnected from the other company. So the thinking on GMs part was that connections to the parent company would stifle innovation and they would reinforce a dominant logic. This is insight number four: Search for potentially synergistic links and cut off those links that limit innovation responsiveness.

It's true, these lessons are easier said than done; but notice their coherence. Synergistic links exploit diversity and positive interdependencies. Selection tempers diversity to help balance exploration and exploitation. Pulling back on efficiency a bit to allow some slack not only ensures robustness, but it promotes greater innovation by allowing diversity. This is true in business and organizations; it's also true in ecosystems, which allow lots of small inefficiencies.

Let's do a little empirical test to see if this passes; just a basic sniff test. Let's go back and look at what sort of organizations have proven robust over time. How can we do that? We can open up a book and we can say, "What organizations have been around not for 5 years, 10 years, 50 years, but 500 years; that's right: 500 years." What are we going to find? We're going to find more than 40 universities—this is going to include Oxford, which is pushing 900—and then only 125 or so businesses. This is surprising, because there are way, way more businesses than universities, and so this suggests universities are far more robust than business. This agrees with what we'd expect from complex systems. Why? Universities have always promoted diversity, and they've always built in a lot of slack to allow for this sort of experimentation. We can sort of understand, then, why so many universities and why so few businesses.

If we look at these 125 or so businesses that have been around for 500 years, guess what? Two-thirds are breweries, taverns, restaurants, and hotels. These have not been businesses historically that try and squeeze out every last nickel; they're sort of laid back business enterprises. The other 40 or so are a mix or candy makers, cheese makers, knife makers, jewelers, and pharmacies. We're not finding banks and construction companies here; we're not finding cutthroat businesses. It turns out a little slack really is a good thing if you want robustness.

Jenna Bednar, a political scientist at the University of Michigan, has written a book called *The Robust Federation*. In that book, she makes the point that this same idea holds for governments. Robust federated governments need slack; they need room for some slippage. The slippage basically allows novel policies that keep governments fresh and responsive. The US Constitution has held us together for more than 200 years, but it's hardly a paragon of efficiency. By permitting freedoms, it encourages diversity; and though it sets incentives in place, they're not that well defined. The Constitution aligns with many of the core insights we've just discussed: be careful how you set incentives; encourage diversity; keep an eye on the tails; don't get too caught up in little efficiency gains; sever unnecessary connections; and encourage synergistic connections.

These may seem straightforward, but I want to stop for just a second and recognize how they depart again from the standard idea of "command and control optimization." If you take a command and control optimization approach, you set incentives with only outcomes in mind. You don't necessarily think about the implications of those incentives on the future set of behaviors and types. You discourage diversity; you want people to be on point. You seek out every possible efficiency gain, and you control the structure of the organization. You don't let people loose to muck things up on the organizational chart. Finally, you make decisions from the top down, not the bottom up. Command and control isn't wrong; it's great if we've hired a group of people to paint a house or build a bridge. But it's not the right thing to do if we're hoping to thrive in a complex world. The world is too complex to be controlled; so try as we might to rationally plan, we're going to wake up to find holes in the ozone layer, CO^2 building up in the atmosphere, bombs going off in public squares, and markets crashing. This is the cost of a complex world. As Emerson wrote, "As soon as there is life there is danger." So be it.

Yet, this same complexity supports emergent phenomena ranging from the sublime structures of snowflakes, which we do understand, to the wondrous consciousness of our own minds, which we don't and may never will. In between these two lay emergent phenomena like the financial networks we described in an earlier lecture. These we may eventually comprehend and we may be able to prevent crashes. Our simple firewall model suggests that appropriate safeguards might emerge with sufficient learning; but empirical evidence suggests that what emerges maybe isn't that robust. Our goal should be to push our understanding of these systems, because only through understanding can we move from a position of poking the tiger with a stick to one where we're taming the lion.

Now that we've reached the end, I want to return to the beginning; or at least the near beginning when we discussed rugged and dancing landscapes. When we confront a fixed rugged landscape, we possess the potential for ultimate success: We solved Fermat's last theorem; we discovered the double helix; and many of us finished the Sunday *New York Times* crossword last weekend. With sufficient time in a rugged landscape, we can map every summit and every valley. But when the landscape dances, we must adapt; we must meet each new challenge with an expanding ensemble of tools and understanding. Our lives play out on dancing landscapes; so we must continue to learn, we must continue to adjust. We wouldn't want it any other way.

Glossary

adaptation: A change in behavior or actions in response to a payoff or fitness function.

agent-based model: A computer model of a complex system that builds from individual agents.

complex adaptive system: A collection of adaptive, diverse, connected entities with interdependent actions.

complicated: A system of connected, diverse, interdependent parts that are not adaptive.

dancing landscapes: Fitness or payoff landscapes that are coupled so that when one entity changes, its action causes the other entity's landscape to shift.

diversity: Differences in the number of types of entities.

emergence: A higher-level phenomenon that arises from the micro-level interactions in a complex system. Emergence can be weak (explicable) or strong (unexplained). Consciousness, for example, is an instance of strong emergence.

explore/exploit: The trade-off between searching for better solutions and taking advantage of what is known.

externality: When an action by one entity influences the payoff or fitness of the actions of another agent. This creates dancing landscapes.

interaction: Effects between the multiple actions of a single entity. These are the cause of rugged landscapes.

interdependence: The influence of one entity's action on the behavior, payoff, or fitness of another entity.

long-tailed distribution: A distribution such as a power law in which most event sizes are small but some are very large.

network: A collection of nodes and links, or connections between those nodes.

nonstationary process: A process in which the probability of events changes over time.

normal distribution: The familiar bell-curve distribution in which most likely event sizes are near the mean.

phase transition: An abrupt change in the macro-level properties of a system.

positive and negative feedbacks: A situation in which an action creates more (positive) or less (negative) of the same action.

power-law network: A network in which the distribution of links fits a model of a type of long-tailed distribution.

robustness: The ability of a complex system to maintain functionality given a disturbance.

rugged landscape: A graphical representation of a difficult problem in which the value of a potential solution is represented as an elevation.

selection: A process through which less fit or lower-performing entities are removed from the population.

self-organization: A form of emergence in which the entities create a pattern or structure from the bottom-up, such as schooling fish.

self-organized criticality: A phenomenon in which interaction agents self-organize into states that can produce large events.

simulated annealing: A search algorithm in which the probability of making an error decreases over time.

small-world network: A network in which the nodes are people and the people have local friends and a few random friends.

tipping point: A configuration in a complex system in which a sequence of events can push the system into a new macro state.

variance: A difference in the value of an attribute.

Bibliography

Anderson, Chris. *The Long Tail: Why the Future of Business is Selling Less of More.* New York: Hyperion, 2006. Anderson describes how new technologies allow for more small niche markets. His long tails are the opposite of the long tails that we discuss in power laws. In power laws, the long tails represent large events. In Anderson's model, the long tail represents many small events.

Anderson, Phillip. "More is Different." *Science* 177 (August 4, 1972): 393–396. Though published in a scientific journal by a Nobel laureate in physics, this paper is accessible and provides one of the earliest and most coherent descriptions of emergence. It is considered one of the founding papers of complex systems theory.

Axelrod, Robert. *The Complexity of Cooperation: Agent-Based Models of Competition and Collaboration.* Princeton, NJ: Princeton University Press, 1997. This book provides some examples of complexity applied to social science problems such as cooperation and the formation of culture. It is an ideal book for someone interested in how social scientists put ideas from complex systems to work by using agent-based models.

Axelrod, Robert, and Michael Cohen. *Harnessing Complexity: Organizational Implications of a Scientific Frontier.* New York: Basic Books, 2001. This book provides the backbone for the final lecture in which I discuss how to harness complexity. Axelrod and Cohen are central figures in complex systems study, and they wrote this book for business people and policy makers to help them understand how to harness the power of complexity.

Bak, Per. *How Nature Works: The Science of Self-Organized Criticality.* 1st ed. New York: Springer, 1996. Here Bak describes his sand-pile model of self-organized criticality in accessible prose. He then shows how the model can be applied to a variety of settings. It is a wonderful book, bursting with Bak's passion for his ideas. This book cannot be described as understated!

Ball, Philip. *Critical Mass: How One Thing Leads to Another.* 1st ed. New York: Farrar, Straus and Giroux, 2004. Ball's book is a wonderful follow-up to Melanie Mitchell's book on complexity. Ball digs much deeper into the physics of complex systems and argues that many of the new ideas of complex systems borrow from existing ideas in physics. He provides a wonderful analysis of phase transitions.

Bednar, Jenna. *The Robust Federation: Principles of Design.* New York: Cambridge University Press, 2008. Most formal models of institutions in social science focus on efficiency: The best institutions are the most efficient. Bednar first shows that efficiency may be impossible in federations owing to imperfect monitoring. She goes on to show that the inherent slippages in a federal arrangement, far from being detrimental, actually enhance robustness by allowing for experimentation and innovation. Her theory also touches on the need for coverage and redundancy in institutional design. All in all, a deep, thoughtful book.

Beinhocker, Eric. *Origin of Wealth: Evolution, Complexity, and the Radical Remaking of Economics.* Cambridge, MA: Harvard University Press, 2007. Written by a leading business consultant and frequent visitor to the Santa Fe Institute, this book describes how complex systems ideas can be used to understand core features of the economy. The book challenges the standard economics orthodoxy.

Epstein, Joshua. *Generative Social Science: Studies in Agent-Based Computational Modeling.* Princeton, NJ: Princeton University Press, 2007. This book contains a collection of articles that rely on agent-based models. Epstein describes how bottom-up agent-based models can be thought of as generative. The book includes both a cogent philosophical contrast between generating an outcome and proving the existence of such an outcome and a wealth of examples that demonstrate the difference.

———. *Growing Artificial Societies: Social Science from the Bottom Up.* Cambridge, MA: MIT Press, 1996. This book describes many of the key concepts in complex systems by using an elaborate agent-based model called Sugarscape. In this model, the agents demonstrate the difference between bottom-up and top-down social science.

Gladwell, Malcolm. *The Tipping Point: How Little Things Can Make a Big Difference*. 1st ed. New York: Little, Brown and Company, 2000. Gladwell shows how very minor adjustments can lead to major impacts in everyday life. This is written for a nontechnical audience; as a result, it is a quick, fun read. Gladwell touches on many of the ideas discussed in the lectures: tipping points, networks, and positive feedbacks among them.

Holland, John. *Adaptation in Natural and Artificial Systems: An Introductory Analysis with Applications to Biology, Control, and Artificial Intelligence*. Cambridge, MA: MIT Press, 1992. This book initiated study in the field of complex adaptive systems. It provides the first full description of genetic algorithms and classifiers and of how systems that adapt can solve problems. This book requires substantial mathematical sophistication on the part of the reader.

————. *Emergence: From Chaos To Order*. New York: Basic Books, 1999. Holland demonstrates that the "emergence" of order from chaos has much to teach us about life, mind, and organizations. This book is written for a more general audience and is a fun, lively read, bursting with ideas. It provides an excellent description of lever points.

————. *Hidden Order: How Adaptation Builds Complexity*. Reading, MA: Helix Books, 1996. Written by one of the founders of complexity theory, this book describes the basic concepts of the theory for a more general audience. Unlike many books on complexity, this book is low on hype and high on substance. For those who want a book with lots of substance (but no equations), this is an ideal choice.

Jackson, Matthew. *Social and Economic Networks*. Princeton, NJ: Princeton University Press, 2008. Jackson's book offers a comprehensive introduction to social and economic networks. Primarily written as a textbook for graduate students in economics, it contains a clear, concise introduction to the theoretical study of networks. Each chapter includes exercises to aid network analysis comprehension.

Kauffman, Stuart. *At Home in the Universe: The Search for Laws of Self Organization and Complexity*. New York: Oxford University Press, 1996. This book focuses on the origins of complexity and self-organization in a biological system. Written for the general science reader, it relies on complex systems ideas to explain how life may have come into being through emergence.

Miller, John, and Scott Page. *Complex Adaptive Systems: An Introduction to Computational Models of Social Life*. Princeton, NJ: Princeton University Press, 2007. This book provides an intermediate-level introduction to complex adaptive systems. It provides formal versions of many of the models discussed in the course and an introduction to agent-based modeling. This book is geared toward the interested general reader with a slight social science bent.

Mitchell, Melanie. *Complexity: A Guided Tour*. New York: Oxford University Press, 2009. An excellent introduction to the field of complex systems written by a leading computer scientist and complex systems scholar. The book draws examples from biology, computer science, and social science. As far as introductions to complex systems go, this is one of the best.

———. *An Introduction to Genetic Algorithms*. Cambridge, MA: MIT Press, 1998. Genetic algorithms, a computer search algorithm introduced by John Holland, encode potential solutions as strings and then use crossover, mutation, and selection to breed new solutions. In this book, Mitchell provides an introduction to genetic algorithms and explains how they can identify and combine partial solutions. This is a great book for anyone interested in learning more about how computer algorithms work.

Newman, Mark. "Power Laws, Pareto Distributions and Zipf's Law." *Contemporary Physics* 46 (September, 2005): 323–351. In this technical academic paper, Newman provides some of the empirical evidence and theories for the existence of power-law forms. This paper requires advanced training in mathematics to understand fully. Nevertheless, as a resource for what might cause power laws, it is unsurpassed.

Newman, Mark, Albert-Lászól Barabasi, and Duncan J. Watts. *The Structure and Dynamics of Networks*. Princeton, NJ: Princeton University Press, 2006. This book provides an overview of the latest breakthroughs in network theory. The book is organized into four sections. It discusses some of the important historical research, the network's empirical side, and the foundational modeling ideas and explores the relationship between network structure and system dynamics.

Page, Scott E. *The Difference: How the Power of Diversity Creates Better Groups, Firms, Schools, and Societies*. Princeton, NJ: Princeton University Press, 2007. This book demonstrates the value of diversity by using models from complex systems. The book shows how rugged landscapes depend on a person's perspective and how collections of bounded agents can produce diverse solutions to problems and make accurate predictions.

———. "Path Dependence." *Quarterly Journal of Political Science* 1 (January 1, 2006): 87–115. In this academic paper, I provide an overview of the causes and types of path dependence. I use a simple model of urns containing two colors of balls to draw distinctions between the types of path dependence. The paper contains a more complete telling of the QWERTY keyboard example.

Raymond, Eric S. *The Cathedral and the Bazaar: Musings on Linux and Open Source by an Accidental Revolutionary*. 1st ed. Sebastopol, CA: O'Reilly, 1999. An excellent book on the difference between top-down organizational structures (the cathedral) and the bottom-up, open-source approach (the bazaar). Raymond's book challenges how many think about organizations and efficiency.

Resnick, Mitchel. *Turtles, Termites, and Traffic Jams: Explorations in Massively Parallel Microworlds*. Cambridge, MA: MIT Press, 1994. Resnick developed a computer program called StarLogo, which was the predecessor to NetLogo, a common agent-based modeling platform. StarLogo was written for younger people to learn to construct agent-based models. Resnick shows how to use StarLogo and agent-based models to produce a variety of phenomena.

Bibliography

Schelling, Thomas. *Micromotives and Macrobehavior*. 1st ed. New York: W. W. Norton, 1978. This book contains the tipping model of segregation discussed in this course. That model is one of many treasures in this tour de force, a book that more than any other was responsible for Schelling winning the Nobel Prize. Many of the phenomena Schelling describes would now be called emergent, though at the time that word was not used in conjunction with the sort of models he describes.

Waldrop, Mitchell. *Complexity: The Emerging Science at the Edge of Order and Chaos*. New York: Simon and Schuster, 1992. This book put the idea of complexity in the public consciousness. It offers up a journalistic account of the history of complex systems research at the Santa Fe Institute. Waldrop describes the big ideas of complex systems through the eyes of the scientists that developed them. It is written as a popular science book that can be read by anyone.

Watts, Duncan. *Six Degrees of Separation: The Science of a Connected Age*. New York: W. W. Norton, 2004. A great introduction to the science of networks. Watts mixes captivating examples with deep theory. The book describes how different network structures exhibit different functionalities, including the famous six degrees of separation.

———. *Six Degrees: The Science of a Connected Age*. New York: W. W. Norton, 2003. This is perhaps the best mass-audience book on networks. Watts writes with clarity and rigor about small-worlds networks, network robustness, and network formation. He peppers his analysis with lively examples.

Weiner, Jonathan. *The Beak of the Finch: A Story of Evolution in Our Time*. New York: Vintage, 1995. This book provides a wonderful description of the work of two biologists who go to the Galapagos and study Darwin's finches. They find that evolution occurs more rapidly than Darwin thought. This book won a Pulitzer Prize.

West, Geoffrey. *Scaling Laws in Biology and Other Complex Systems*. Google Tech Talks, 2007; 54 min., 30 sec.; streaming video, http://video.google.com/videoplay?docid=7108406426776765294. This is a Google TechTalk that examines how universal scaling laws follow from fundamental principles and lead to a general quantitative theory that captures essential features of many diverse biological systems. The talk is somewhat technical but provides a powerful demonstration of the fan-out nature of complex systems.

Notes

Notes